Collins

KS2 Maths Reasoning

SATs 10 Minute Tests

Katherine Pate

How to Use this Book

This book contains 14 Key Stage 2 Reasoning tests, each designed to be completed in approximately 10 minutes.

Consisting of SATs-style questions in bite-sized chunks, each 10-minute test will help children to prepare for the SATs Reasoning paper at home.

Clearly laid out questions and easy-to-use answers will help your child become familiar with, and gain confidence in, answering and understanding SATs-style questions.

The first two tests are intended to be a little easier, as an introduction to SATs practice. However, the tests can be done in any order and at any time throughout Year 6 to provide invaluable practice for your child.

- Children should work in a quiet environment where they can complete each test undisturbed. They should complete each test in approximately 10 minutes.
- The number of marks available for each question is given on the right-hand side of the test pages, with a total provided at the end of each test.
- Some questions have a 'Show your method' box. For these questions, you may get a mark for showing your method, even if your answer is incorrect.
- Answers and marking guidance are provided for each test.
- A score chart can be found at the back of the book, which your child can use to record their marks and see their progress.

Acknowledgements

The author and publisher are grateful to the copyright holders for permission to use quoted materials and images.

Images are © Shutterstock.com and © HarperCollins*Publishers*

Every effort has been made to trace copyright holders and obtain their permission for the use of copyright material. The author and publisher will gladly receive information enabling them to rectify any error or omission in subsequent editions. All facts are correct at time of going to press.

Published by Collins
An imprint of HarperCollins*Publishers*
1 London Bridge Street
London SE1 9GF

HarperCollins*Publishers*
Macken House, 39/40 Mayor Street Upper,
Dublin 1, D01 C9W8, Ireland

ISBN: 9780008335892

Content first published 2019
This edition published 2020
Previously published by Letts

10 9

British Library Cataloguing in Publication Data.

A CIP record of this book is available from the British Library.

Author: Katherine Pate
Commissioning Editors: Michelle l'Anson and Fiona McGlade
Editor and Project Manager: Katie Galloway
Cover Design: Sarah Duxbury and Kevin Robbins
Inside Concept Design: Ian Wrigley
Text Design and Layout: Jouve India Private Limited
Production: Karen Nulty
Printed in India by Multivista Global Pvt. Ltd

This book contains FSC™ certified paper and other controlled sources to ensure responsible forest management.

For more information visit: www.harpercollins.co.uk/green

Contents

Test 1

1 Tick the time shown on the clock.

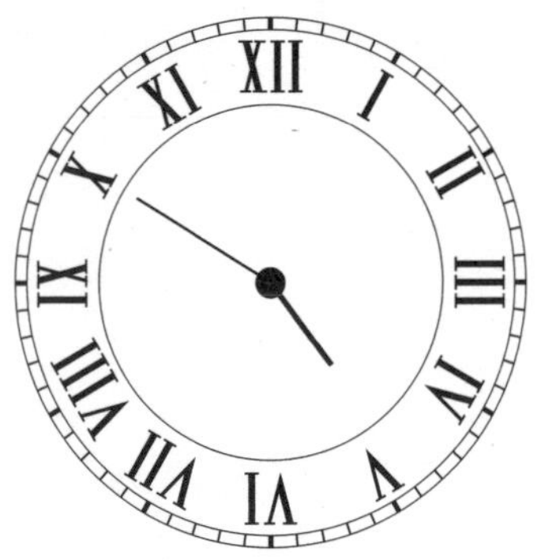

Tick **one**.

Ten to four ☐

Five to four ☐

Ten to five ☐

Ten to six ☐

Twenty five past ten ☐

1 mark

2 The numbers in this sequence increase by the same amount each time.

Write the missing numbers.

☐ 48,234 49,234 ☐ 51,234 ☐

2 marks

3 Draw the lines of symmetry on this rectangle.

2 marks

4 The table shows the numbers of people working in different factories.

Factory	Population
Factory A	6,214
Factory B	15,976
Factory C	23,249

How many **more** people work in Factory C than in Factory A and Factory B **combined**?

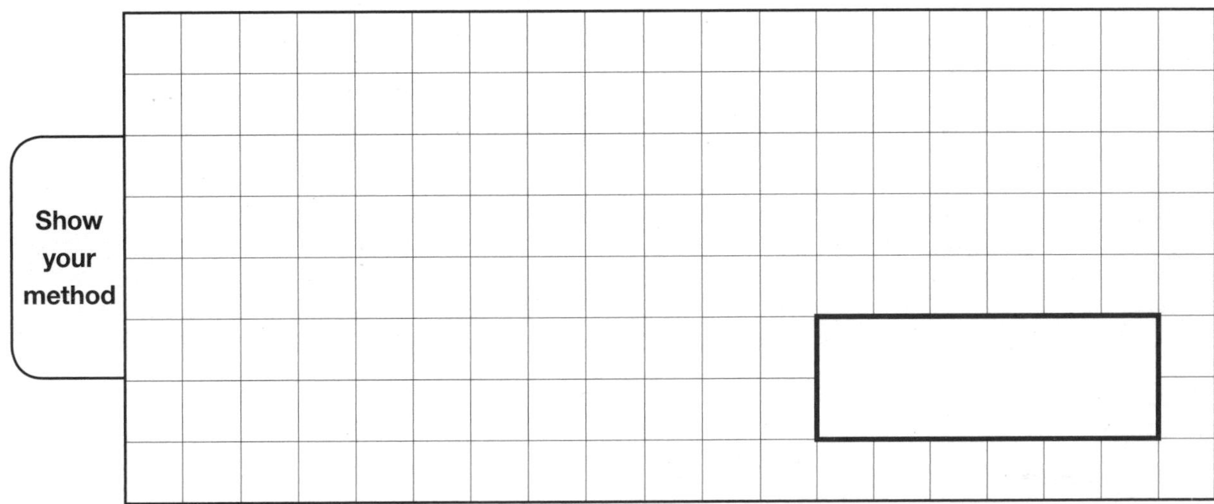

2 marks

5 The thermometers show the temperature at 5pm and 10pm.

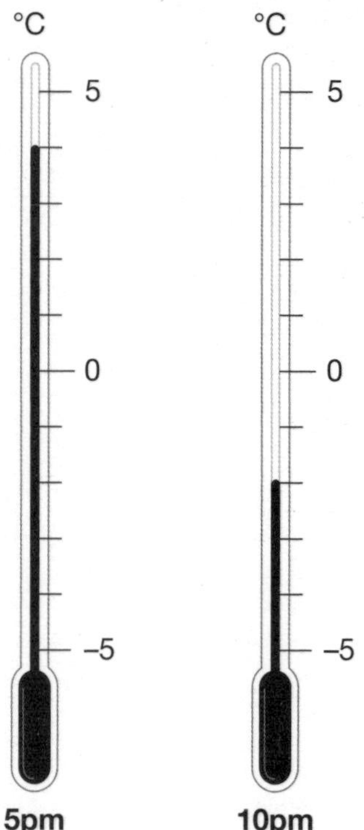

How many degrees did the temperature **fall** between 5pm and 10pm?

.... °C

1 mark

At midnight it was 2 degrees **colder** than at 10pm.

What was the temperature at midnight?

.... °C

1 mark

6 In this circle, $\frac{1}{3}$ and $\frac{1}{2}$ are shaded.

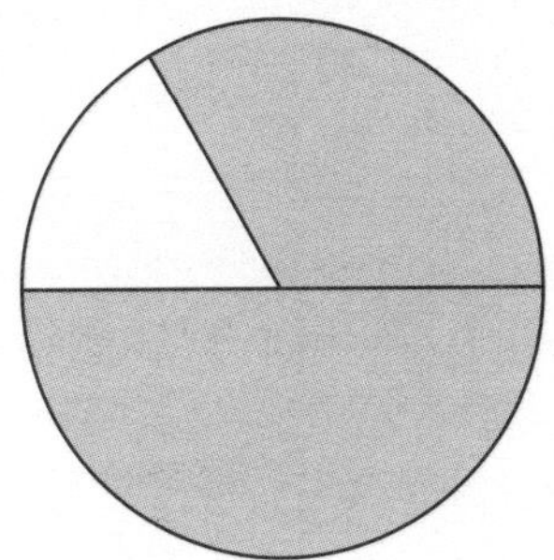

What fraction of the whole circle is **not** shaded?

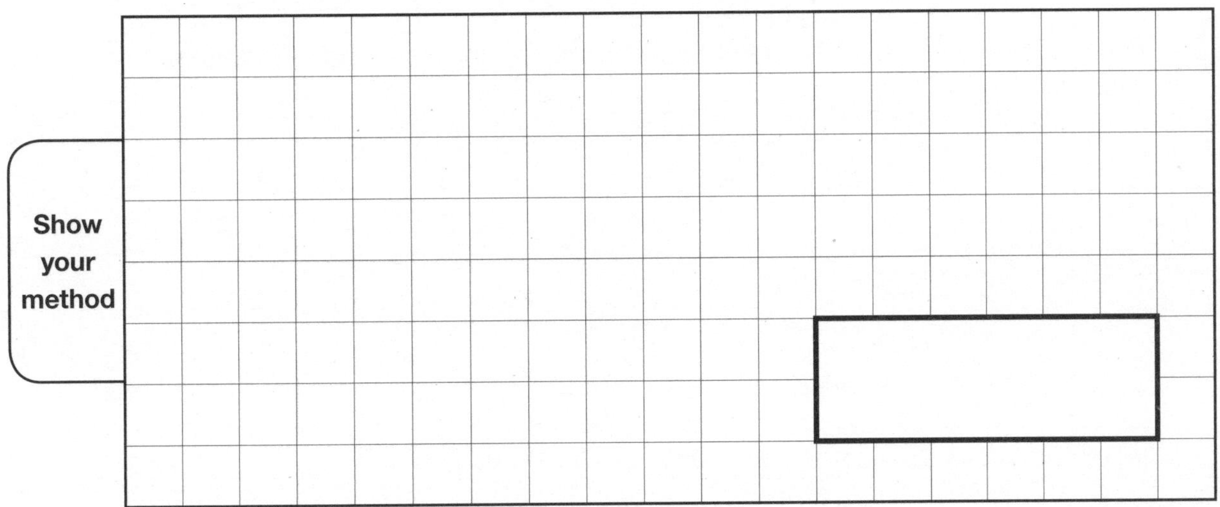

2 marks

Work out the angle for the unshaded sector.

....

1 mark

Test 2

1 How many months have **more** than 30 days?

Tick **one**.

Exactly five months have more than 30 days. ☐

Exactly six months have more than 30 days. ☐

Exactly seven months have more than 30 days. ☐

Exactly eight months have more than 30 days. ☐

1 mark

2 A field is a rectangle with sides measuring 50 metres and 35 metres.

50 metres

35 metres

Lily runs once around the perimeter of the field.

Calculate how far Lily runs.

Show your method

2 marks

3 A shape is translated from position A to position B.

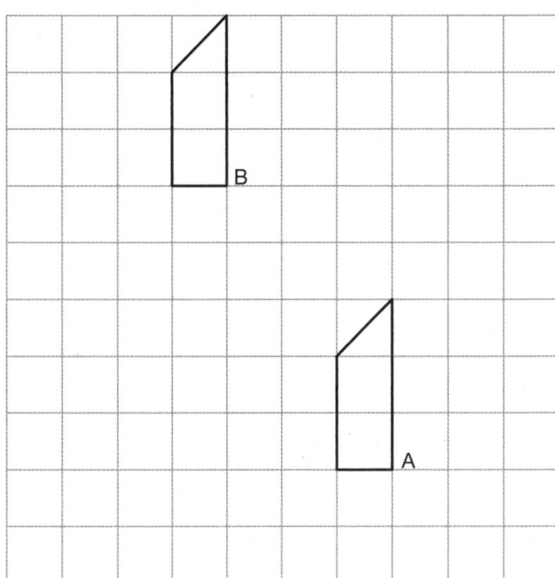

Write the missing numbers in the boxes.

The shape has moved ☐ squares to the left and ☐ squares up.

1 mark

4 A square number and a prime number both less than 50 have a difference of 11.

What are the two numbers?

☐ – ☐ = 11

square number – **prime number**

1 mark

5 On a video game, Mark's score is 23,050 points.

He loses 2,480 points.

Then he wins 8,040 points.

Calculate his new score.

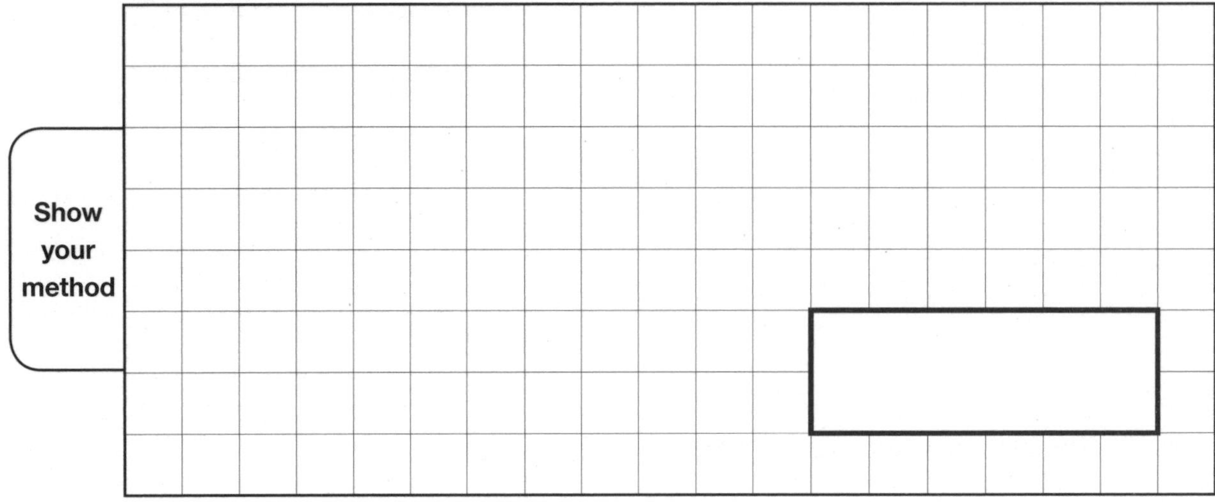

2 marks

6 Sara bought five lollies.

Four of the lollies cost £1 each.

The other lolly cost £2.

What was the **mean** cost of the lollies?

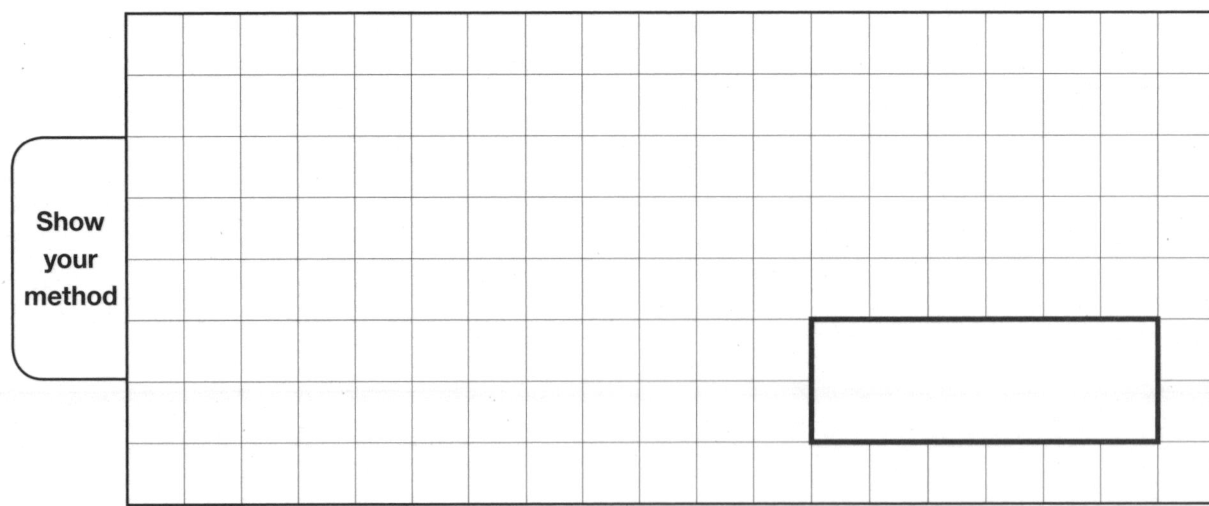

2 marks

7 4 packets of cookies cost £19.88

Each packet contains 14 cookies.

Calculate the cost of one cookie, to the nearest penny.

Show your method

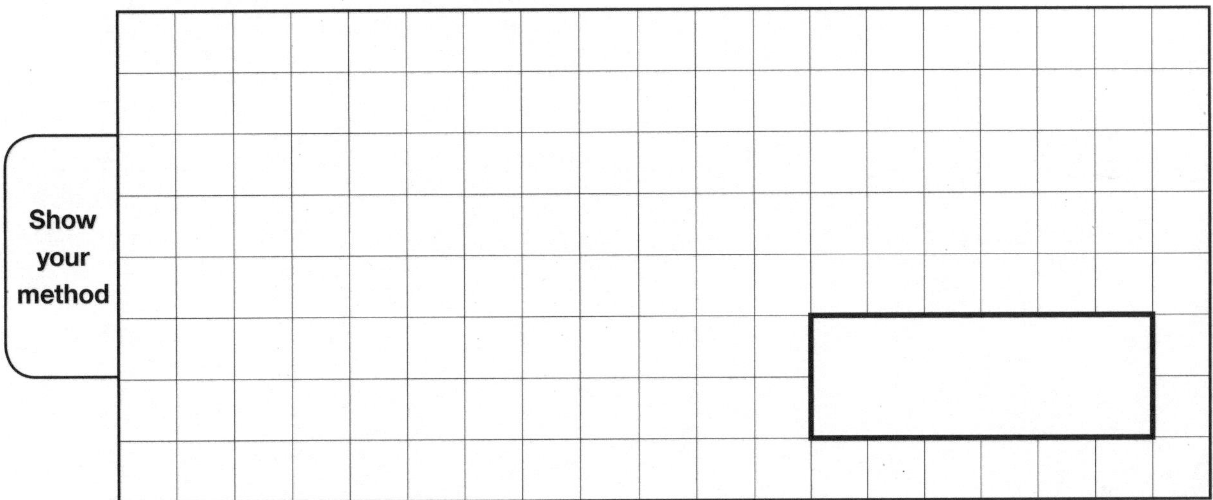

3 marks

Test 3

1 The pictogram shows the colours of cars made in one year.

Red	●	◐	
Black	●	●	
Silver	●	●	●
Other	●		

Key: ● represents 1,000 cars.

How many more silver cars are there than red cars?

☐ cars

2 marks

2 Complete the table.

	Round 24,387
to the nearest 10	
to the nearest 100	
to the nearest 1,000	

2 marks

3 The numbers in this sequence decrease by the same amount each time.

$\frac{9}{11}, \frac{7}{11}, \frac{5}{11}, \frac{3}{11}, \ldots$

What is the next number in the sequence?

1 mark

4 A car drives 16 km on 1 litre of petrol.

The car has 37 litres of petrol in its petrol tank.

Does the car have enough petrol to drive 500 km?

You must show your calculation.

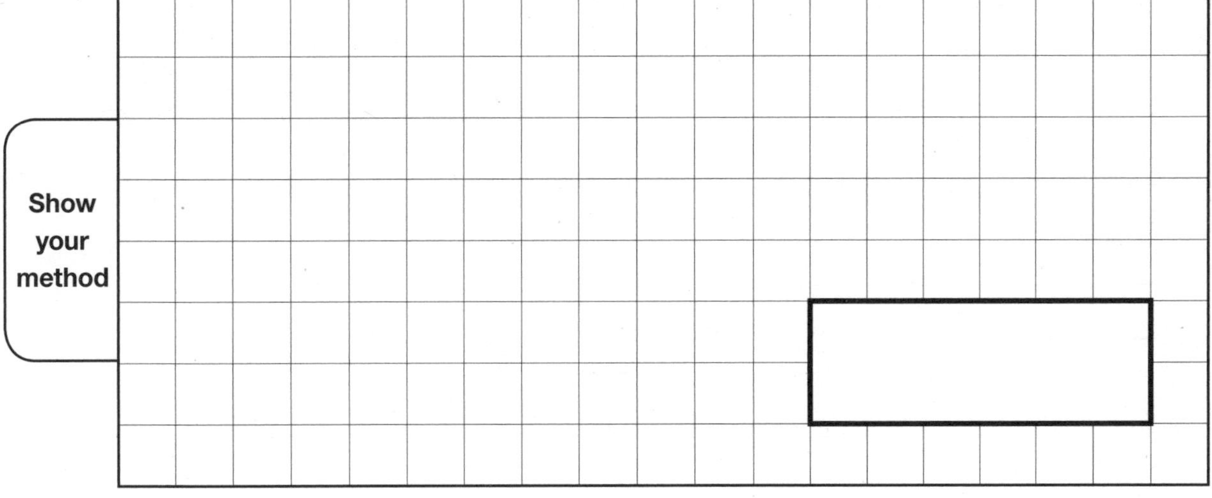

2 marks

5 This shape is made from two rectangles.

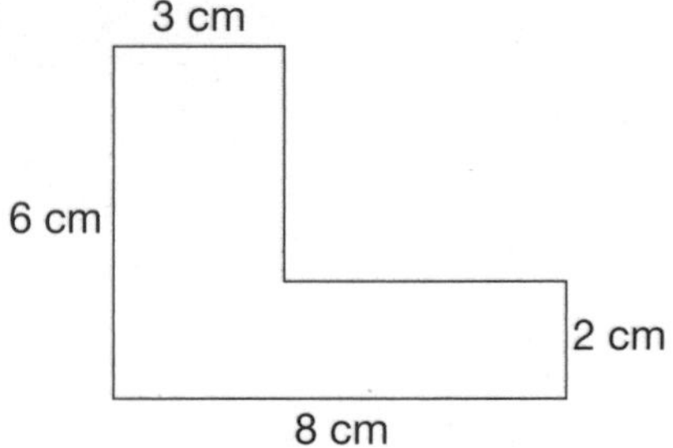

Calculate the **perimeter** of this shape.

[ ] cm

3 marks

6 A shop ices names on chocolate bars.

They use this formula to work out the price for a chocolate bar with a name on it.

price = 20p × number of letters + £1.50

What is the price of a chocolate bar with a 5-letter name on it?

£ [ ]

1 mark

10 min

Test 4

1 In this grid, there are four multiplications.

Write the **three** missing numbers.

4	×	12	=	
×		×		
7	×		=	42
=		=		
		72		

1 mark

2 On this pentagon:

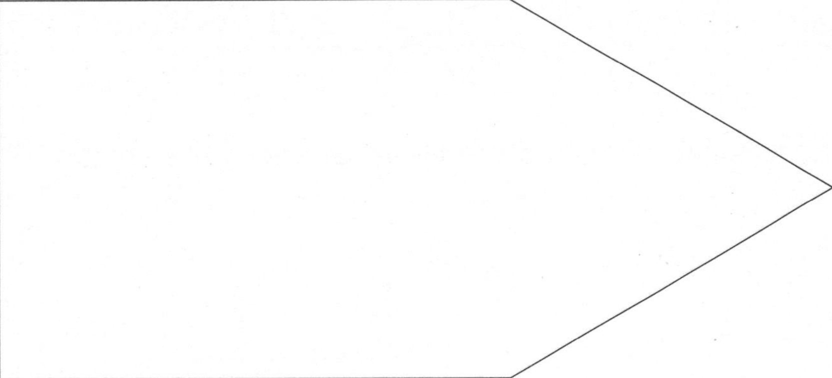

Label a right angle with an R.

Label an acute angle with an A.

Label an obtuse angle with an O.

2 marks

3 Leah uses these three cards to make a 3-digit number.

8	7	3

She divides her 3-digit number by 100.

The answer has 8 hundredths.

What could Leah's 3-digit number be?

1 mark

4 This is a bus timetable from Grant Street to the station.

Grant Street	0852	1139	1526
Bute Road	0857	1144	1531
Hill Street	0904	1151	1538
Station	0910	1157	1544

How long does the bus journey take from Grant Street to the station?

☐ minutes

1 mark

Between which two stops is the longest part of the bus journey?

Between .... and ....

1 mark

5 Order the numbers, starting with the **smallest**.

Match each number with its order.

75%
$\frac{4}{5}$
0.4
$\frac{7}{10}$
$\frac{1}{2}$

1st	smallest
2nd	
3rd	
4th	
5th	largest

1 mark

6 A square tile measures 15 cm by 15 cm

A rectangular tile is 2 cm **wider** and 4 cm **shorter** than the square tile.

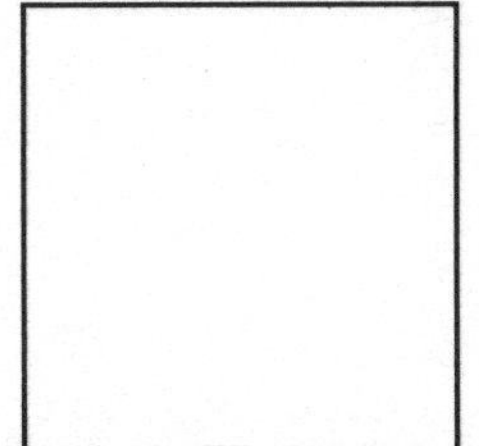

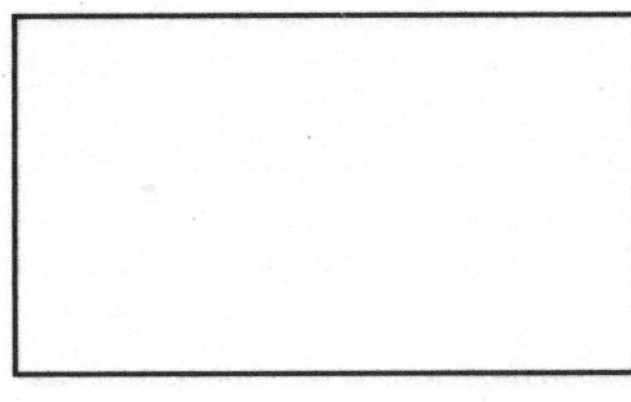

What is the difference in the perimeter of the two tiles?

Show your method

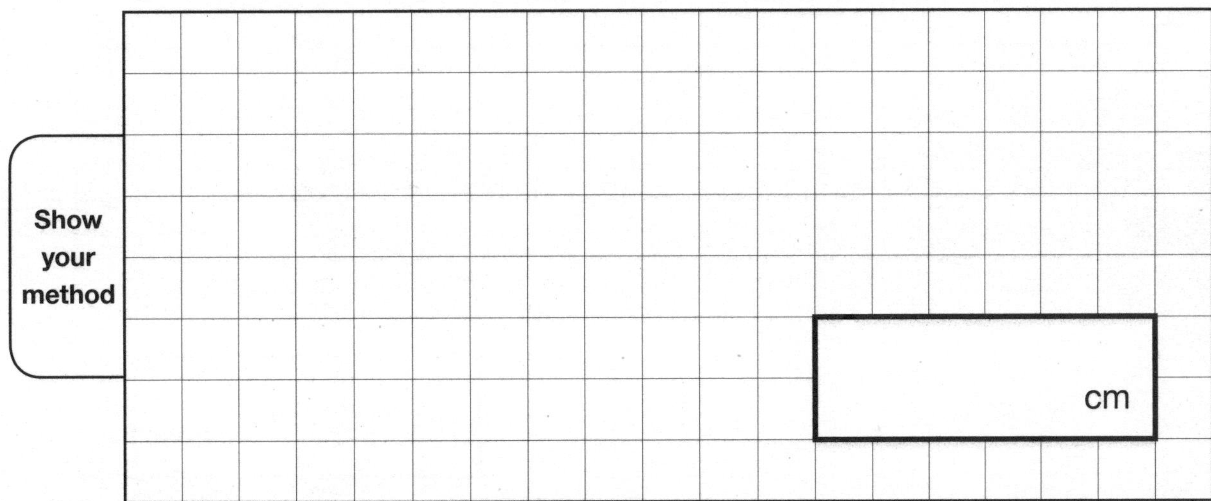

3 marks

Test 5

1 The numbers in this sequence decrease by the same amount each time.

Write the missing numbers.

.... 976 956 .... 916 ....

2 marks

2 There are 4 cans of beans in a pack.

There are 8 packs of cans of beans in a box.

How many cans of beans are there in 7 boxes?

Show your method

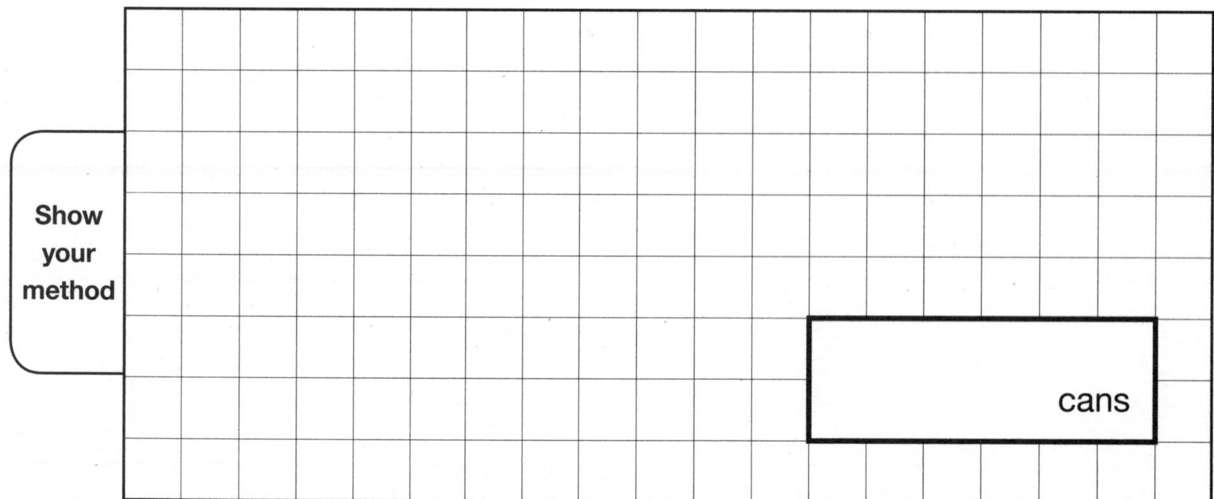

2 marks

3 Tim draws a square on this coordinate grid.

Three of the vertices are marked.

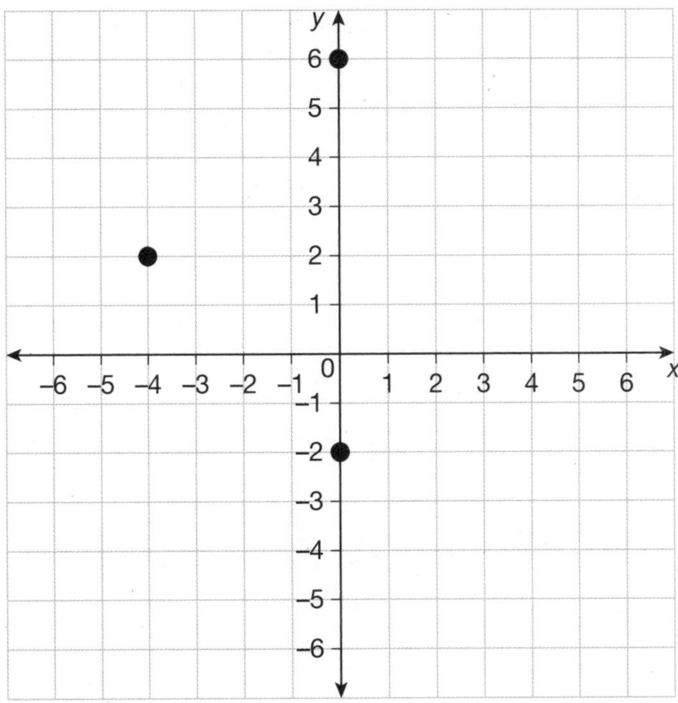

What are the coordinates of the missing vertex?

(.............., ...)

1 mark

4 The average January temperature in Moscow is –6°C.

The average January temperature in Oslo is 7° higher.

What is the average January temperature in Oslo?

 °C

1 mark

5 Ben has 9 pizzas delivered to his party.

Each pizza costs £6.99

The delivery charge is 50p per pizza, plus an extra £5

How much does Ben pay altogether?

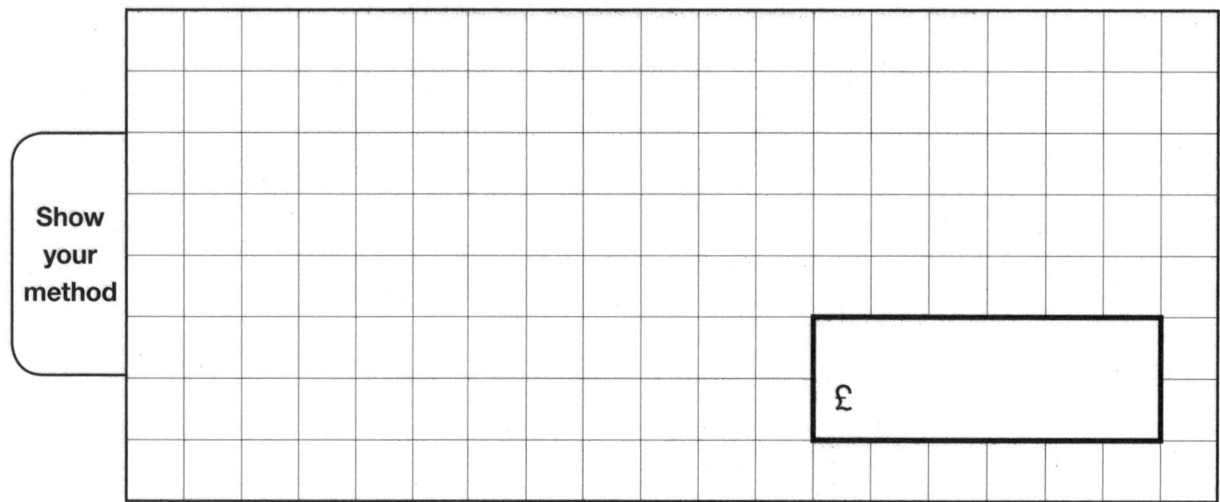

3 marks

6 The diameter of Pluto is 0.18 times the diameter of the Earth.

The diameter of the Earth is approximately 12,740 km

Which is the best estimate for the diameter of Pluto?

Tick **one**.

130 km ☐

260 km ☐

2,600 km ☐

13,000 km ☐

26,000 km ☐

1 mark

Test 6

1 Write these masses in order, smallest first.

2000 g 3 kg 200 g 30 g

.................., ..., ..., ...

2 marks

2 Here are five shapes on a square grid.

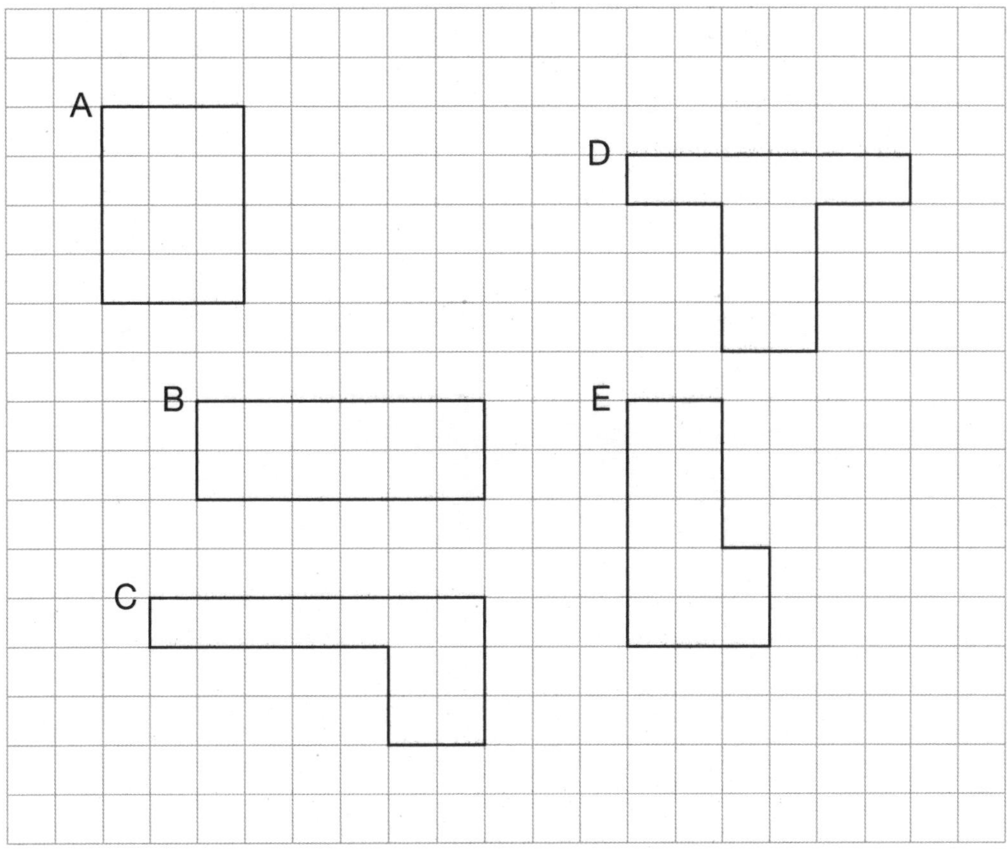

Four of the shapes have the same area.

Which shape has a **different** area?

Shape ☐

1 mark

3 Write the missing number to make this **division** correct.

134 ÷ ☐ = 1.34

1 mark

4 Write three prime factors of 30.

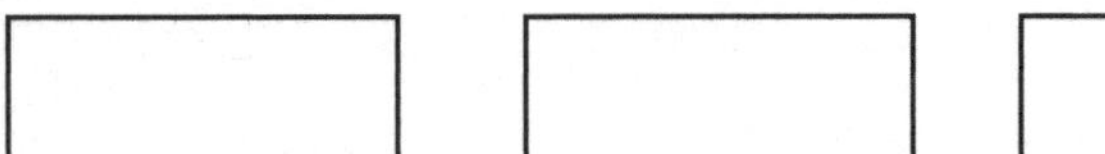

2 marks

5 Chloe and Bella are both reading the same book.

The book has 312 pages.

Bella has read $\frac{1}{4}$ of the book.

Chloe has read $\frac{1}{3}$ of the same book.

How many more pages has Chloe read than Bella?

Show your method

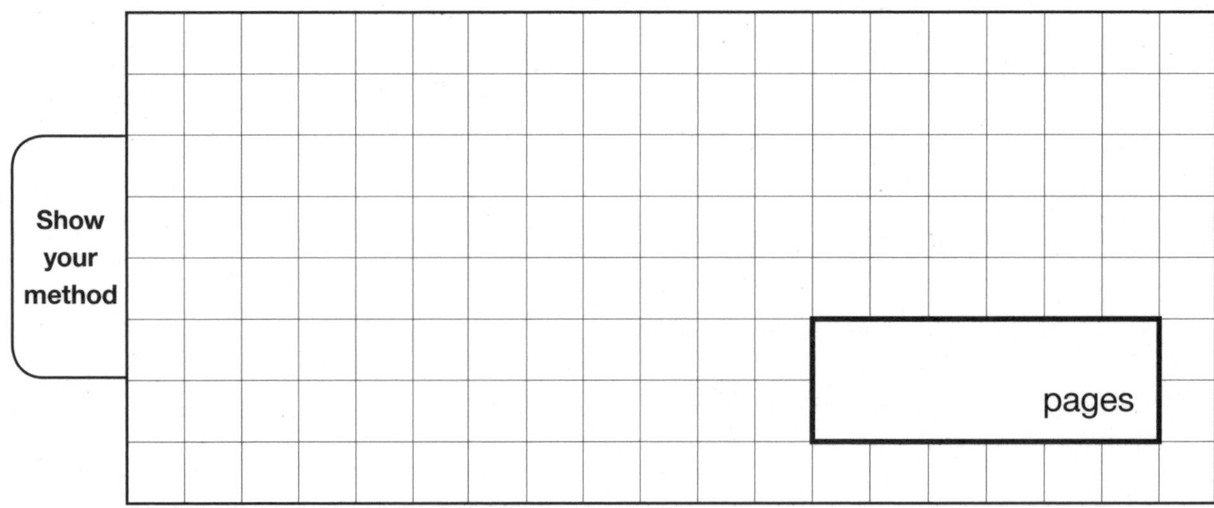

3 marks

6 Draw the triangle with vertices at (–2, –1), (0, 4) and (3, –2) on this coordinate grid.

Label your triangle A.

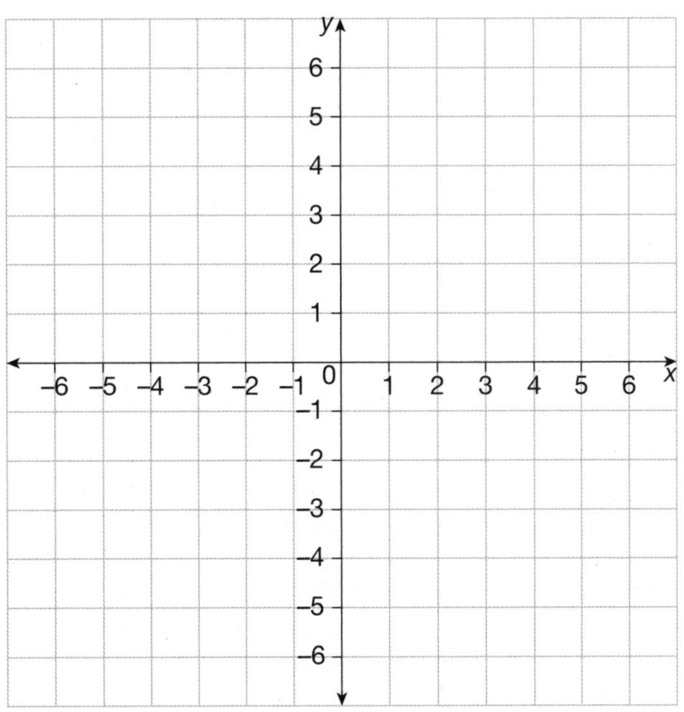

1 mark

Triangle A is translated **4 left** and **3 down**.

Draw the triangle in its new position.

Label this triangle B.

1 mark

Test 7

1 Circle the shape with all angles **smaller than** a right angle.

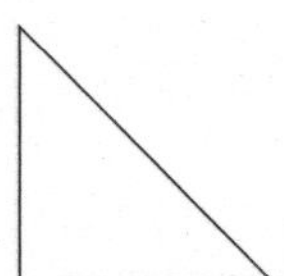
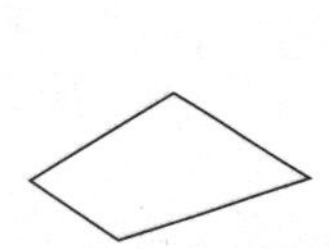
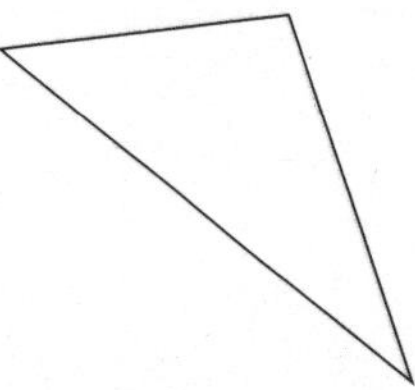
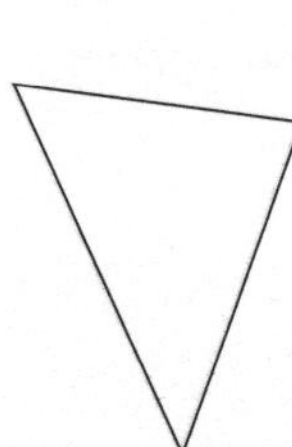
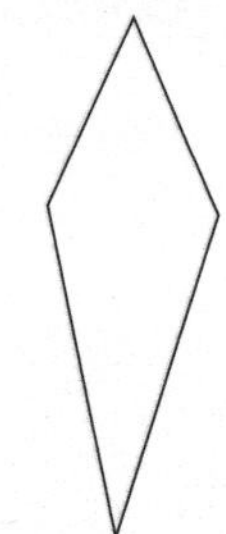

1 mark

2 Kim had £542 in her bank account at the start of the week.

During the week she

- paid £1,740 into her bank account
- took £987 out of her bank account.

Work out how much she had in her bank account at the end of the week.

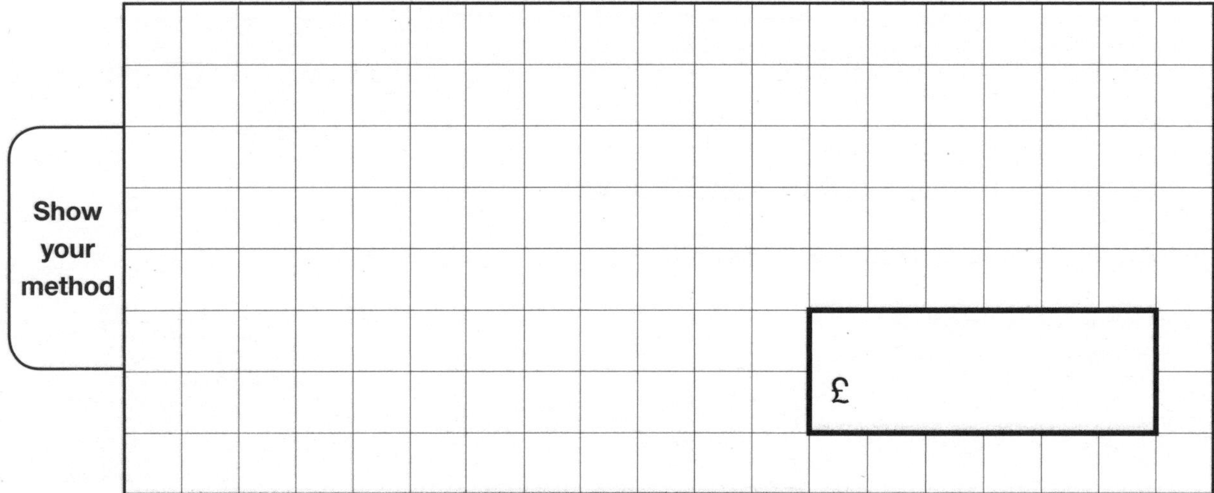

2 marks

3 Circle the improper fraction that is equivalent to $4\frac{7}{9}$

$\frac{36}{9}$ $\frac{43}{9}$ $\frac{45}{9}$ $\frac{47}{9}$ $\frac{74}{9}$

1 mark

4 This diagram shows a shaded shape and two lines of symmetry.

Draw the rest of the shape.

Use a ruler.

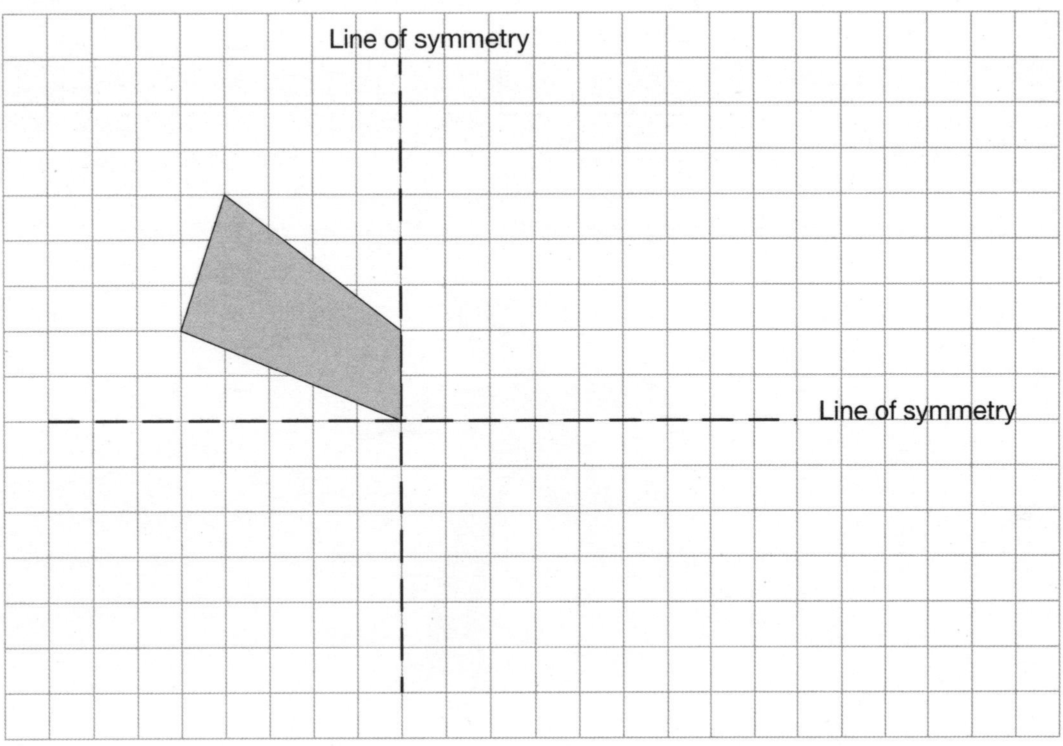

1 mark

5 A factory makes 63,750 chocolate bars an hour.

How many chocolate bars does it make in 20 minutes?

1 mark

6 Round 234,819

to the nearest 10

to the nearest 100

to the nearest 1,000

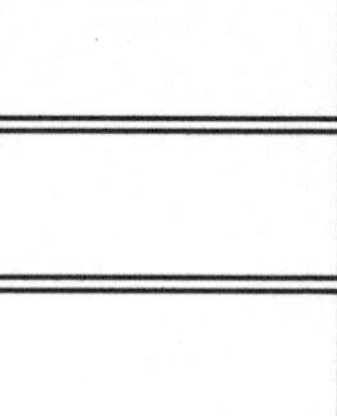

2 marks

7 Last year Jack had three holidays: in Spain, England and Mexico.

The mean cost of his three holidays was £1,150

His holiday in Spain cost £680

His holiday in Mexico cost £2,350

How much did his holiday in England cost?

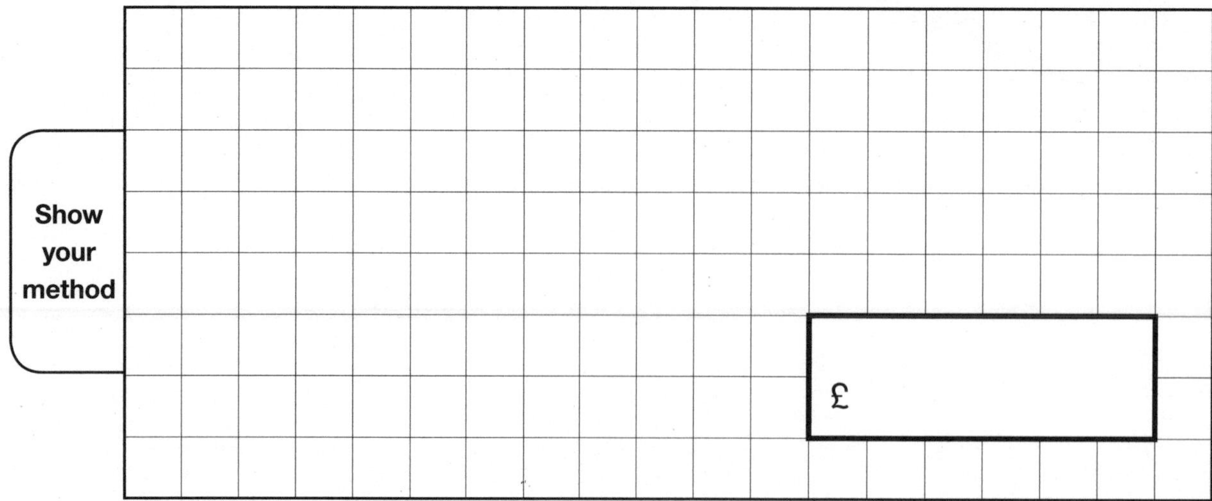

3 marks

Test 8

1 Lou completes this calculation.

$$\begin{array}{r} 38 \\ +\ 43 \\ \hline 81 \\ \hline \end{array}$$

Write a subtraction calculation she could use to check her answer.

	□	□
–	□	□
	□	□

1 mark

2 Write the coordinates of the point marked.

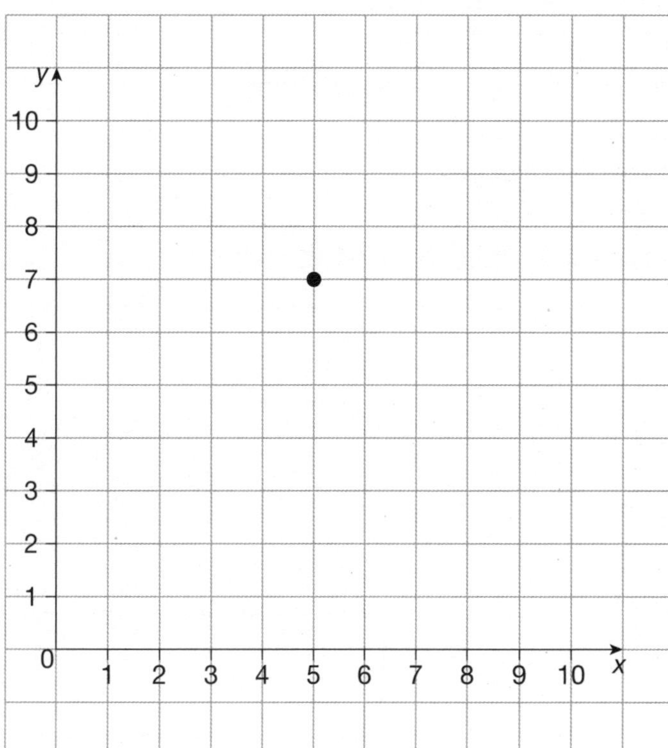

(............., ...)

1 mark

3 Write each number in its correct place in the diagram.

1 4 8 16

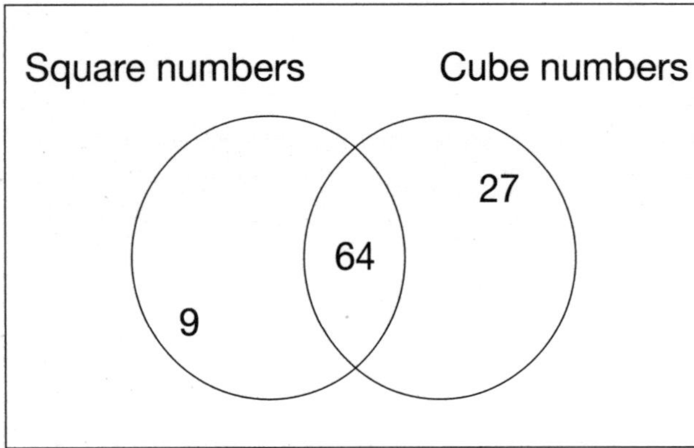

2 marks

4 Mel and Jim run children's parties.

Mel uses this formula to work out the cost for her parties:

Cost of party = £8 × number of children + £35

Jim uses this formula to work out the cost for his parties:

Cost of party = £12 × number of children

Who will run the cheapest party for 7 children – Mel or Jim?

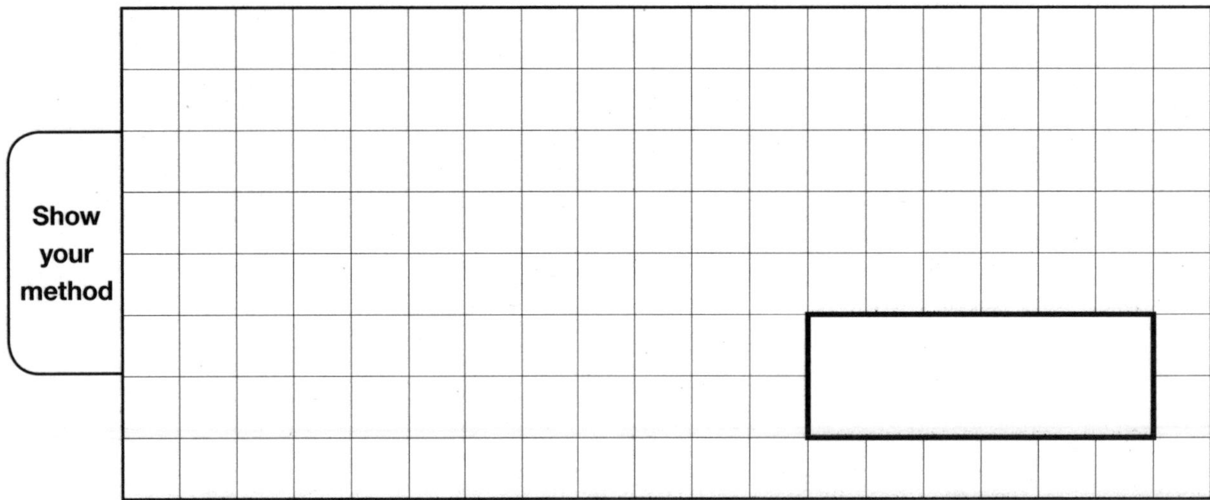

2 marks

5 Write 0.48 as a fraction in its simplest form.

....

2 marks

6 Cube A has sides 3 cm long.

Cube B has sides 1 cm longer than cube A.

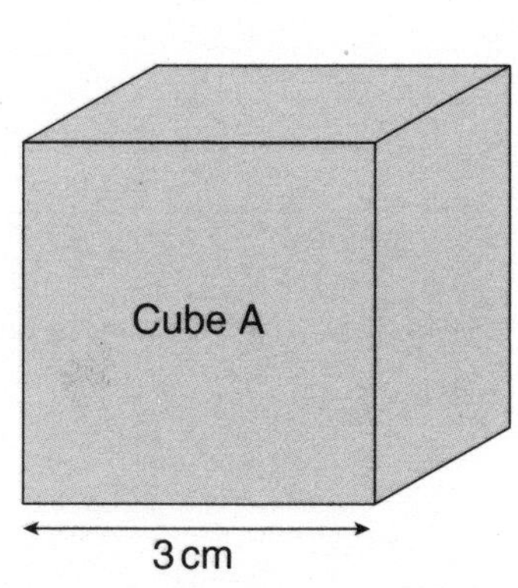

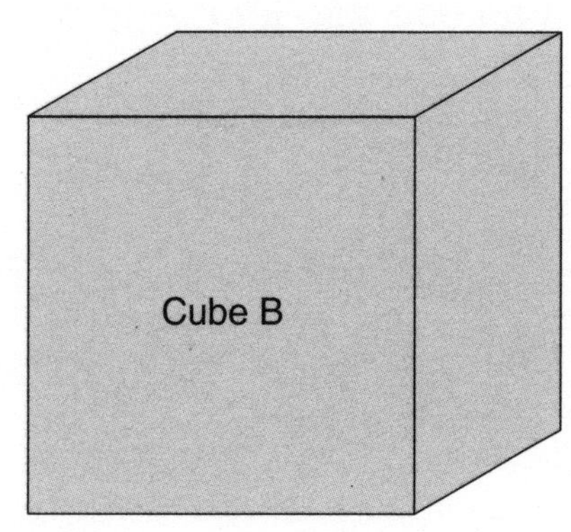

What is the difference in volume between the two cubes?

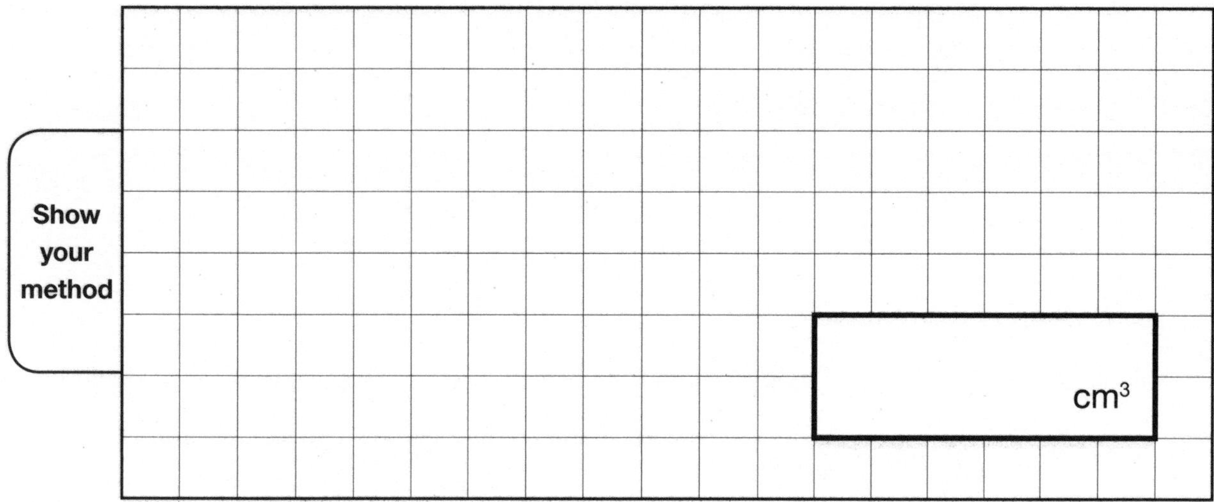

3 marks

Test 9

1 Jake puts these five fractions in their correct places on a number line.

$\frac{1}{6}$ $\frac{1}{2}$ $\frac{1}{4}$ $\frac{1}{11}$ $\frac{1}{5}$

Write the fraction closest to 1 ☐ 1 mark

Write the fraction closest to 0 ☐ 1 mark

2 13 × 51 = 663

Explain how you can use this fact to find the answer to 51 × 14

1 mark

3 Amina and Dawn buy some hair clips.

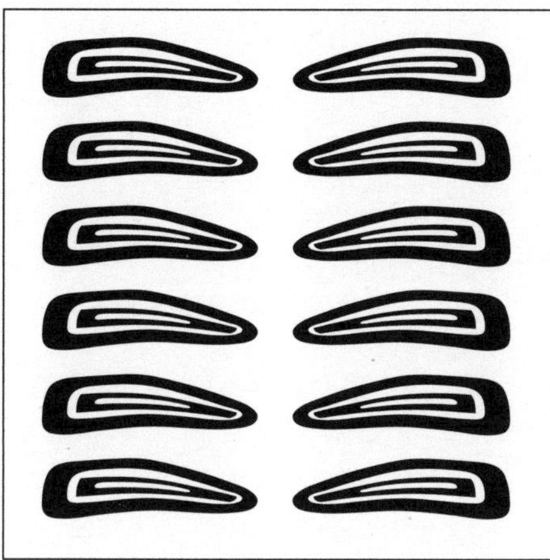

Pack of 12 hair clips £2.99

12 hair clips 29p each

Amina buys a pack of 12 hair clips for £2.99

Dawn buys 12 hair clips for 29p each.

How much more does Dawn pay than Amina?

Show your method

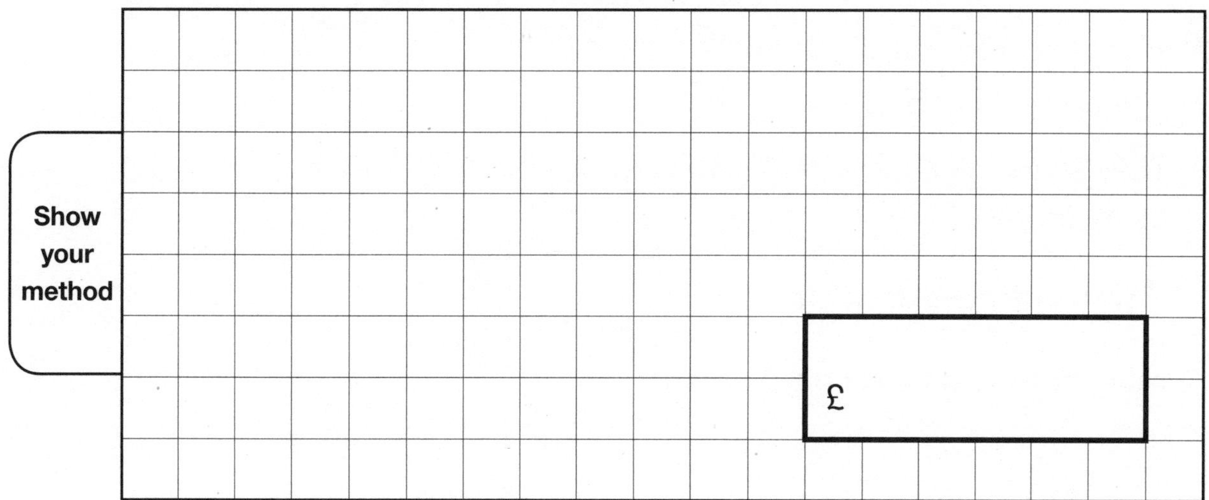

£

2 marks

4 This is a 1 cm cube.

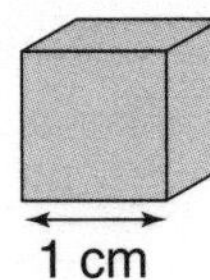

How many of these 1 cm cubes will fit in this cuboid?

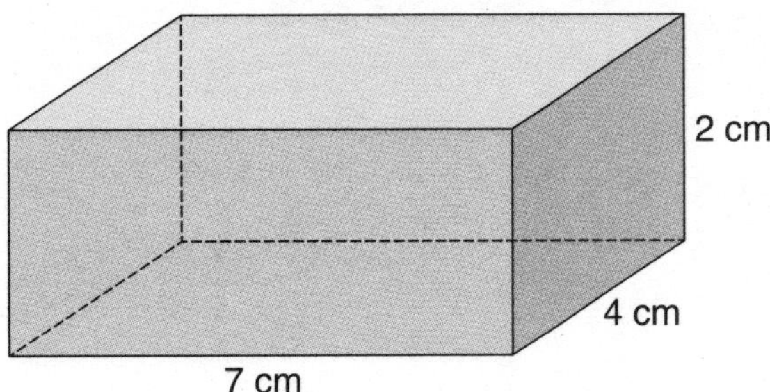

1 mark

5 Jack and Lucy buy their mother a present.

The present costs £21

They share the cost so that Jack pays twice as much as Lucy.

How much does Jack pay? £ 1 mark

How much does Lucy pay? £ 1 mark

6 In this diagram, ABC is an isosceles triangle with side AB = side AC.

Calculate the sizes of angles x, y and z in this diagram.

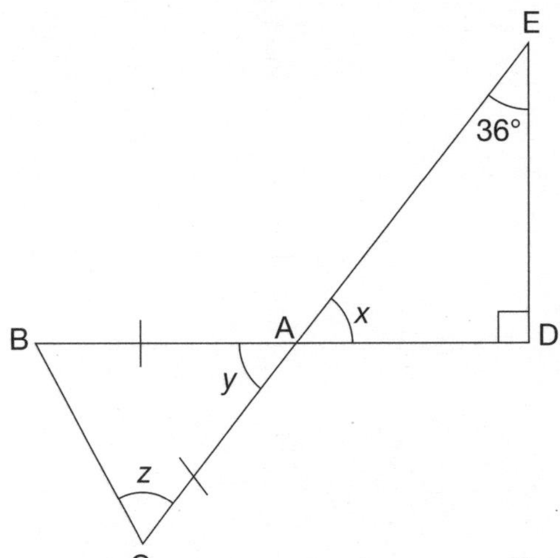

Not to scale

$x =$ ☐ ° 1 mark

$y =$ ☐ ° 1 mark

$z =$ ☐ ° 1 mark

Test 10

1 Leila buys a cake.

She pays with a **£5 note**.

This is her change:

What is the cost of the cake?

£ ☐

1 mark

2 Write the two missing digits to make this **multiplication** correct.

	4	7	☐
×			8
3	☐	7	6

2 marks

3 This graph shows Rani's height every year from birth to 7 years old.

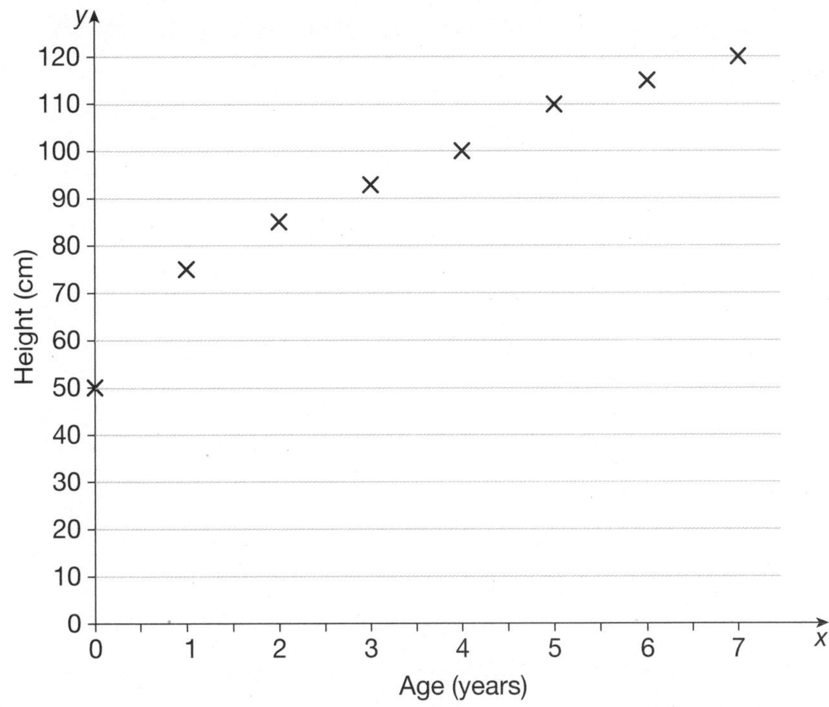

How many cm **greater** was her height at age 7 than at age 5?

[ ] cm

1 mark

At what age was her height double her height at birth?

[ ] years old

1 mark

4 The length of a day on Earth is 24 hours.

The length of a day on Neptune is $\frac{2}{3}$ of the length of a day on Earth.

What is the length of a day on Neptune?

[ ] hours

1 mark

5 Draw the reflection of this triangle in the *y*-axis.

Use a ruler.

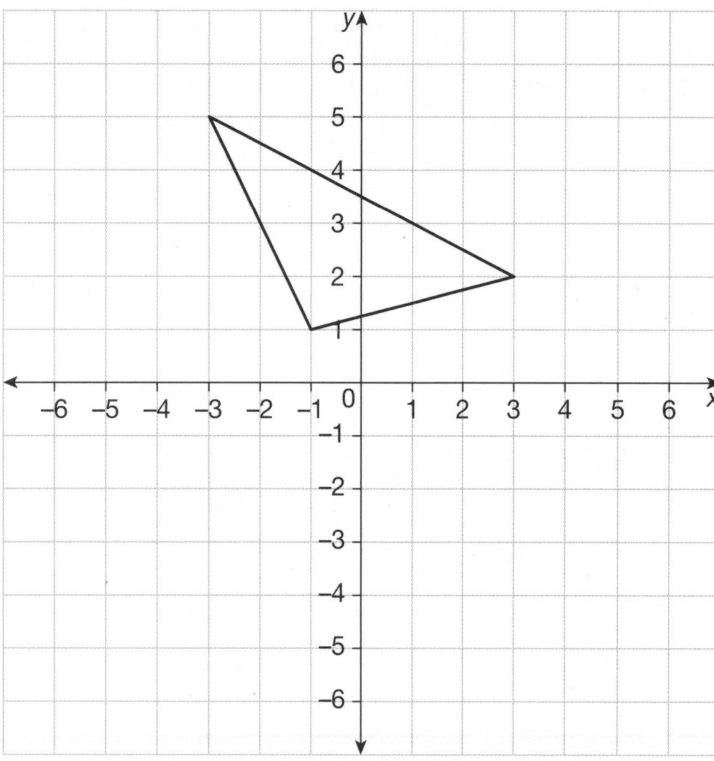

1 mark

6 A number of people were asked their favourite colour.

The pie chart and the table show the results.

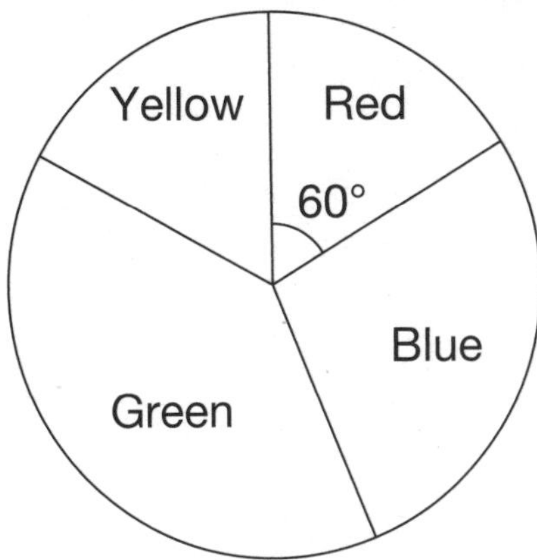

Write the missing values in the table.

Angle	Favourite colour	Number of people
60°	Red	12
100°	Blue	....
..........°	Green	26
..........°	Yellow	....
	Total	....

3 marks

Test 11

1 These diagrams show three equivalent fractions:

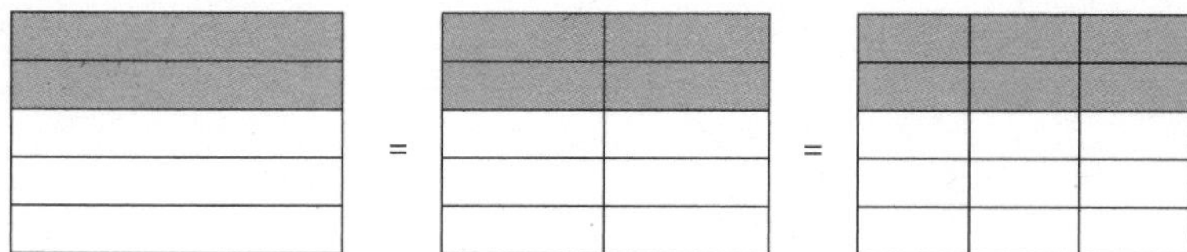

Write the missing values.

$$\frac{2}{5} = \frac{4}{\square} = \frac{\square}{15}$$

1 mark

2 Write each shape name in the correct place in the table.

rectangle **square** **parallelogram** **kite**

Shape	Number of lines of symmetry
....	0
....	1
....	2
....	4

2 marks

3 The date a building was opened is given in Roman numerals.

Write the year MCMLIII in figures.

1 mark

4 This glass has vertical sides.

The diameter of its base is 68 mm.

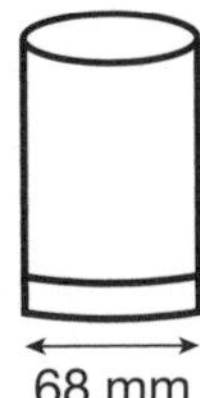

How many of these glasses will fit on the rectangular tray?

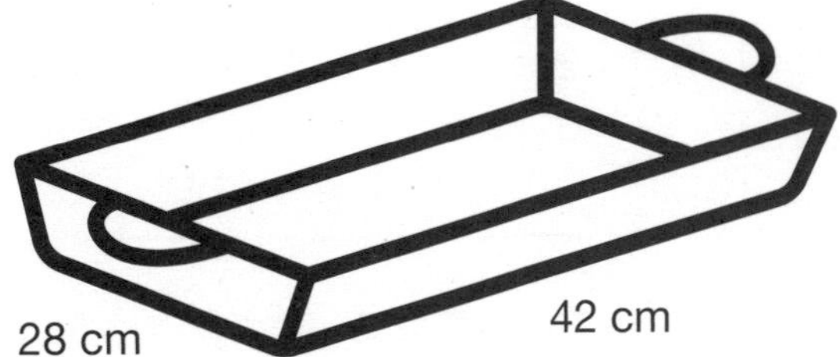

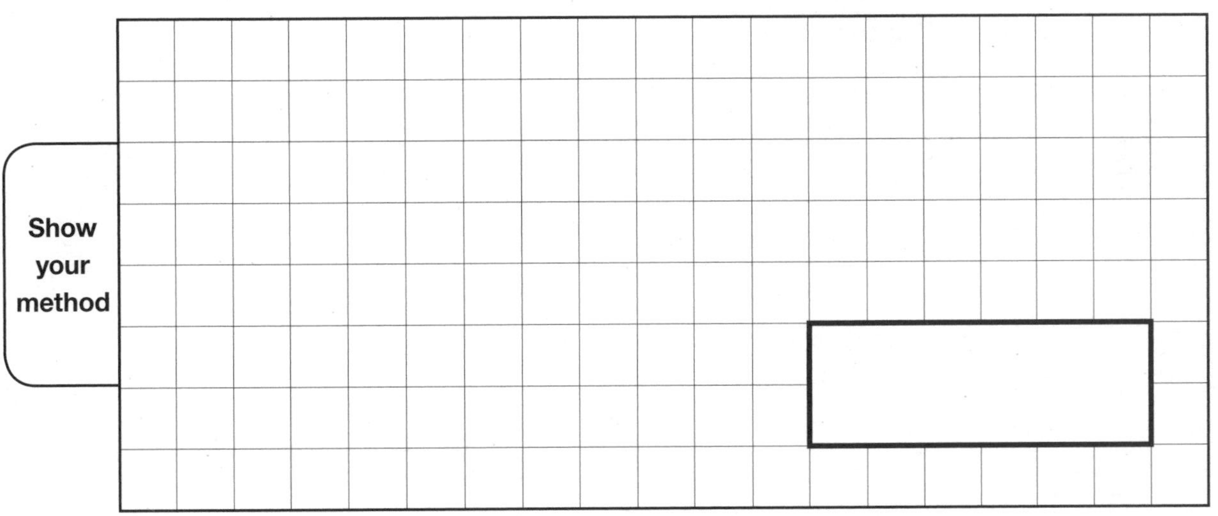

2 marks

5 Here are five numbers.

51 52 53 ~~54~~ 56

Write each number on the correct cards.

The number 54 has been written on the correct cards for you.

Prime numbers	Multiples of 2	Multiples of 3
	54	54

2 marks

6 Here is the bill for a meal for 3 friends.

3 main meals	£24.90
3 desserts	£15
2 colas	£2.40
1 fruit juice	£1.70
Total:	£44.00

The 3 friends add a 10% tip.

Then they share the bill equally.

How much should each friend pay?

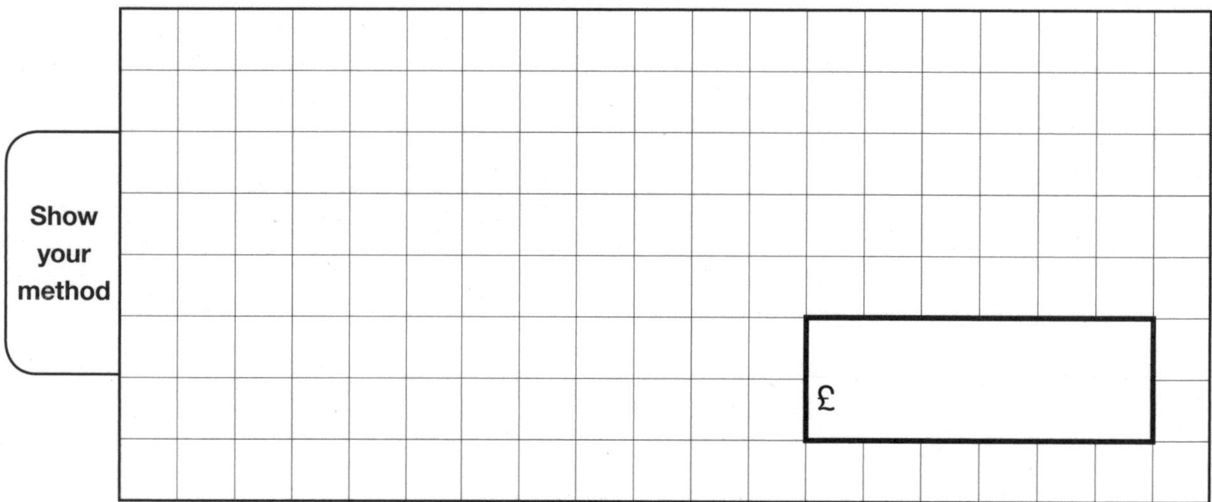

3 marks

Total marks .... /11

10 min

Test 12

1 Write this number **in figures**:

Three thousand, two hundred and seven

1 mark

Write **in words**, the number that is:

10 less than 5,004

1 mark

2 Rita and Kaylum share a pizza.

Rita eats $\frac{1}{9}$ of the pizza.

Kaylum eats 5 times as much as Rita.

What fraction of the pizza is left over?

1 mark

3 Circle two numbers that add together to make 1.

0.237   0.327   0.583   0.673   0.783

1 mark

4 This is a plan of a hall floor.

7.2 m

12 m

Joe is varnishing the floor.

One tin of varnish

- costs £15
- is enough for 8 m^2

Work out the cost of varnish for the hall floor.

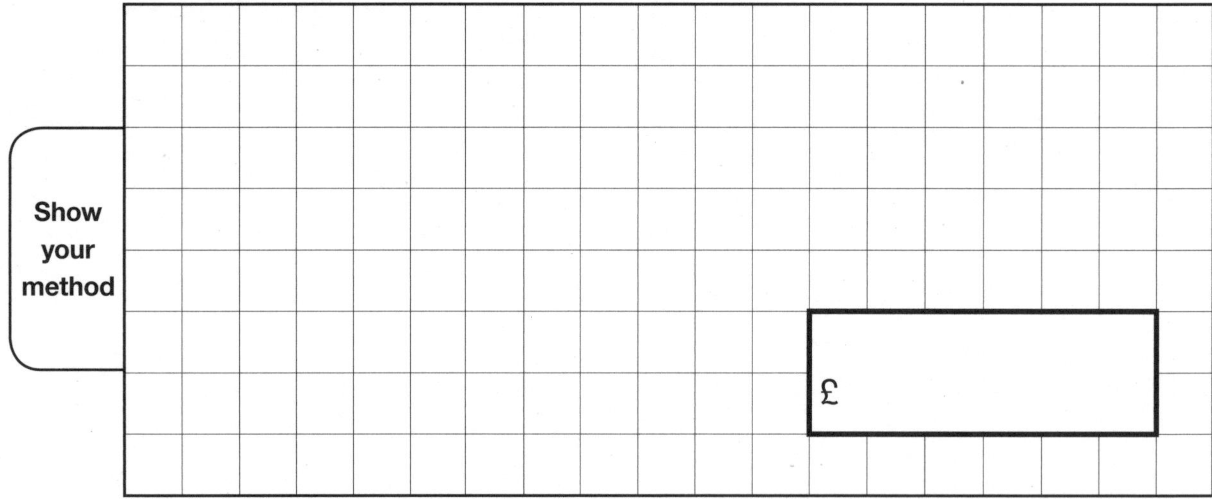

3 marks

5 $\boldsymbol{n = 7}$

What is $5n - 2$?

1 mark

$\boldsymbol{3t + 2 = 20}$

Work out the value of t $\boldsymbol{t} =$

1 mark

6 This is a recipe to make pancakes for 4 people.

100 g plain flour
2 eggs
300 ml milk
15 g butter

Mary has 1 litre of milk.

Does Mary have enough milk to make pancakes for 10 people?

You must show your calculation.

Show your method

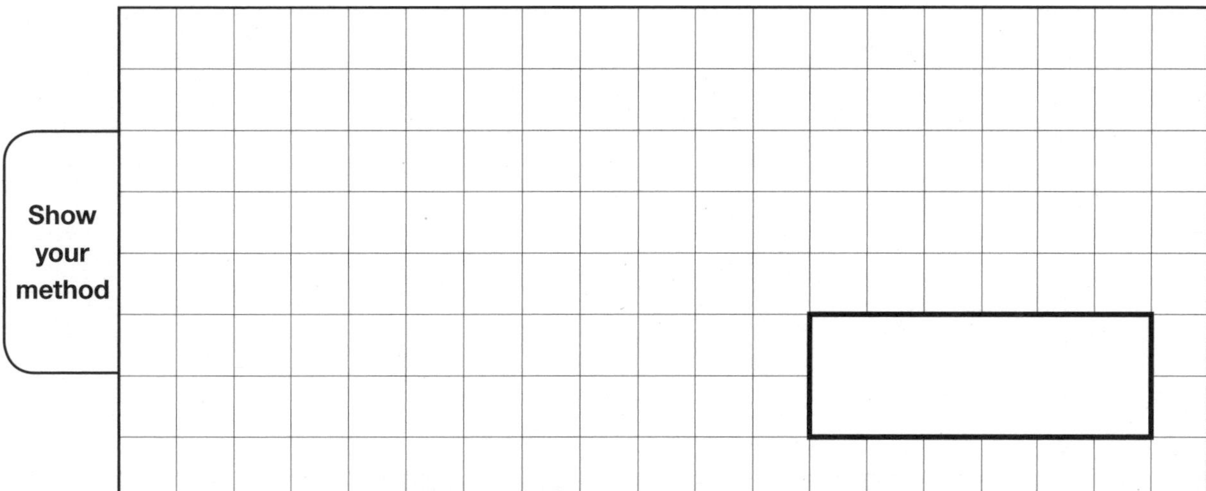

2 marks

Test 13

1 Sara chooses two **different** flavours of ice cream.

There are four flavours to choose from: **vanilla, mint, toffee, lemon**.

Write the two missing combinations:

The ice cream could be:

- vanilla and mint
- vanilla and toffee
- vanilla and lemon
- mint and toffee
- .... and ....
- .... and ....

1 mark

2 Write the missing number to make this **multiplication** correct.

$2.07 \times \square = 207$

1 mark

3 Write the correct sign, <, = or > between these two numbers.

$\frac{2}{5}$ ☐ 25%

1 mark

4 Complete this table with the missing numbers.

One row has been done for you.

Number	10,000 more
84,000	94,000
165	....
243,000	....
....	14,750

2 marks

5 Here are five shapes on a square grid.

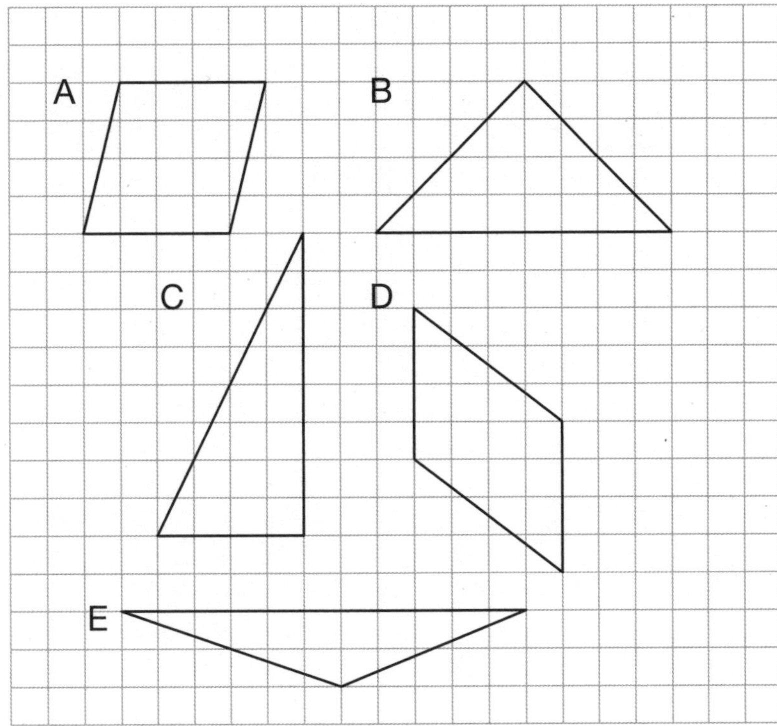

Four of the shapes have the same area.

Which shape has a **different** area?

Shape

1 mark

6 Here are some nets.

Tick (✓) the net that will **not** make a cube.

Tick **one.**

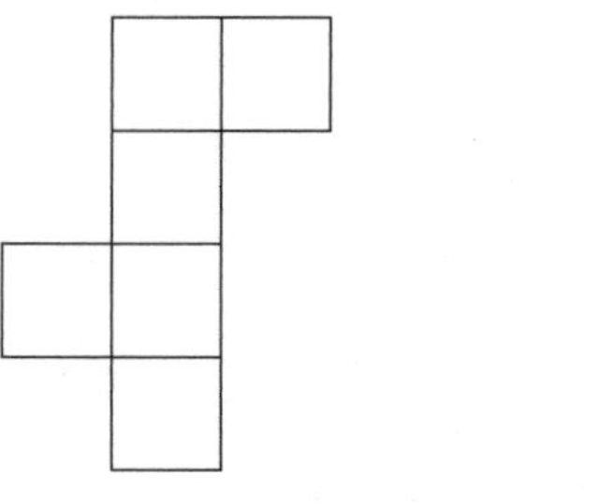
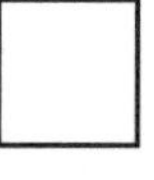

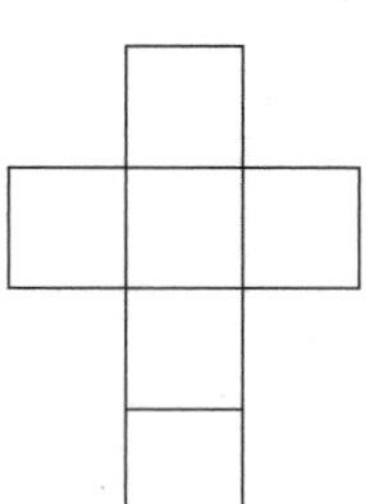
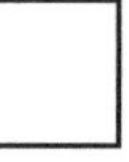

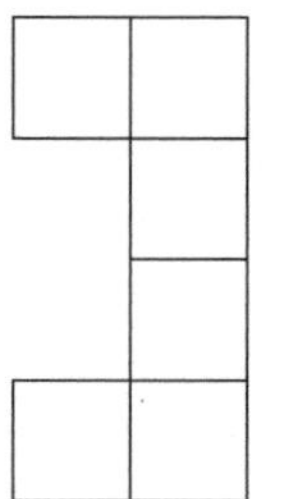
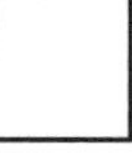

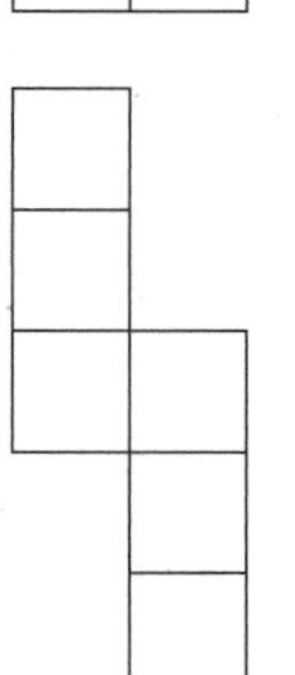
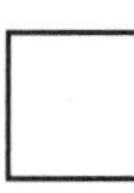

1 mark

7 Here is a rule for the cooking time for a chicken.

Cooking time = 40 minutes for each kilogram + an extra 20 minutes.

How many minutes will it take to cook a 2 kg chicken?

[ ] minutes

1 mark

Deb buys a 3 kg chicken for dinner.

Dinner is at 6.30 pm

What is the latest time she can start cooking the chicken?

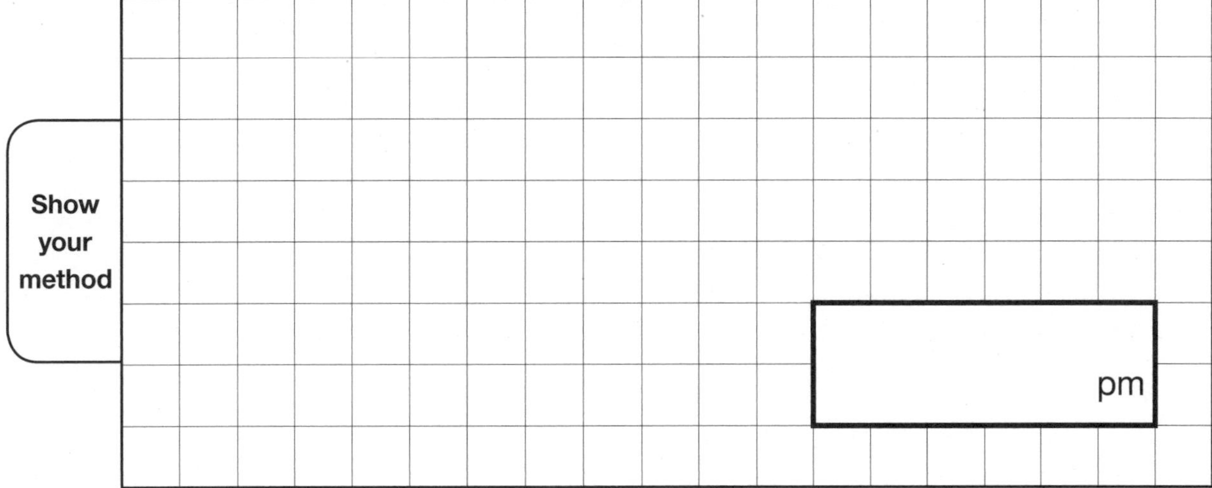

2 marks

Test 14

1 The table shows the numbers of patients a doctor sees each day one week.

Day	Number of adult patients	Number of child patients
Monday	14	37
Tuesday	9	25
Wednesday	10	22
Thursday	14	18
Friday	23	16

Altogether, how many adult and child patients did the doctor see on Wednesday?

1 mark

How many **more** children did the doctor see on Monday than on Friday?

1 mark

2 $73 \times 8 = 584$

Use this fact to work out:

$73 \times 80 =$ ☐

1 mark

$73 \times$ ☐ $= 58.4$

1 mark

3 Ali's race time was 1 minute 40 seconds.

Lily's race time was 10% longer.

What was Lily's race time?

☐ minute ☐ seconds

1 mark

4 Two of the angles in a triangle are 25° and 65°.

Jayden says:

Explain why he is correct.

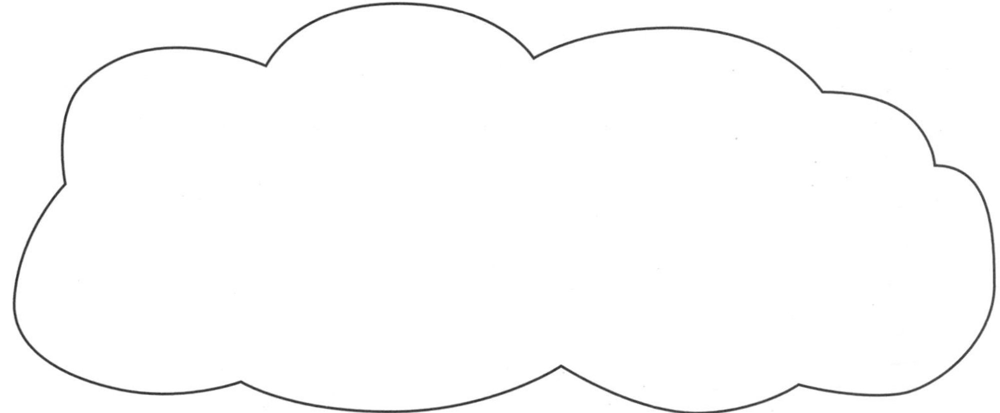

1 mark

5 Freya makes designs with two shapes of tiles.

The total cost of tiles in this design is £30:

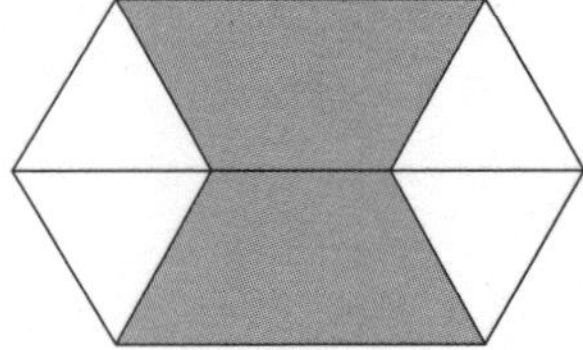

The total cost of tiles in this design is £38:

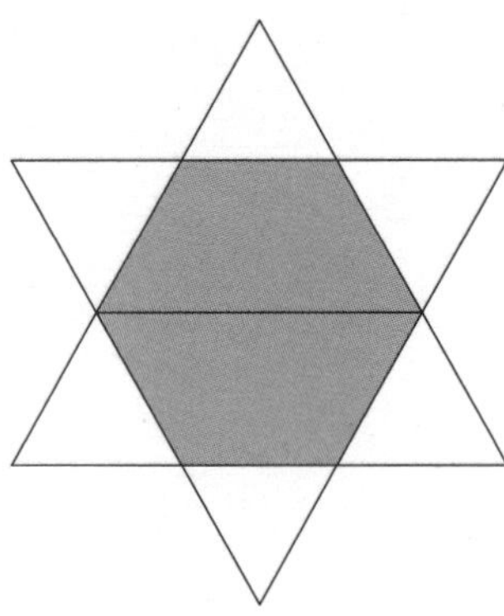

Calculate the cost of each shape of tile.

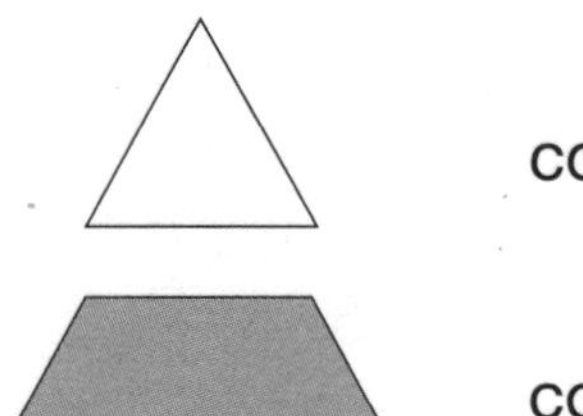

costs £ ______ 1 mark

costs £ ______ 1 mark

6 A wall in a library is 5.4 metres long.

Bookcases are 960 mm wide.

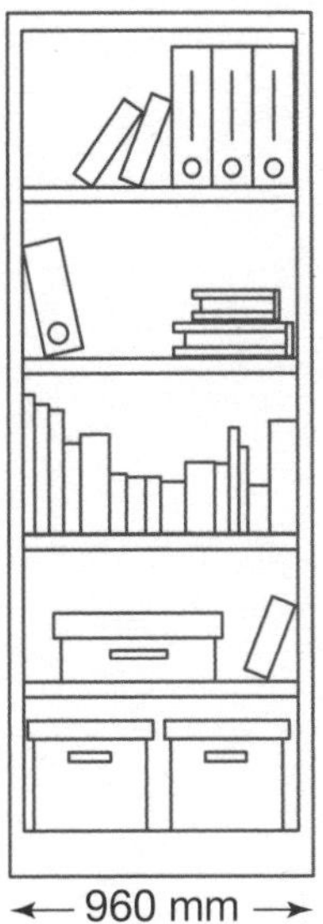

How many bookcases will fit along the wall?

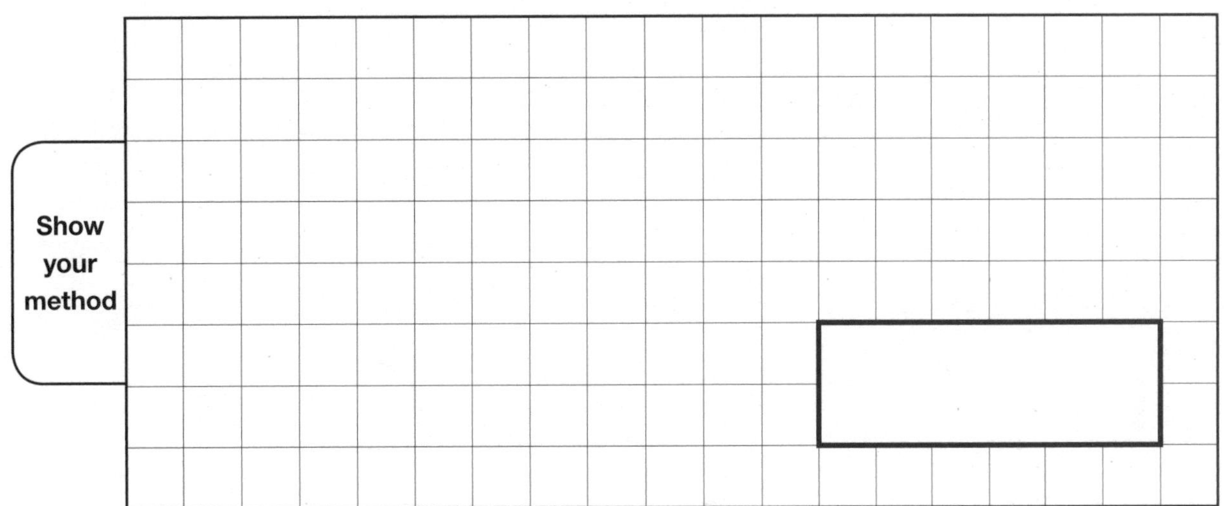

2 marks

How much space is left over? mm

1 mark

Notes

Question	Answer	Mark	Additional guidance
Test 1			
1	Ten to five ✓	1	
2	Award **TWO** marks for three correct numbers: 47,234 50,234 52,234 Award **ONE** mark for two of these numbers correct.	Up to 2	
3	Award **TWO** marks for two correct lines of symmetry as shown (and no other lines): 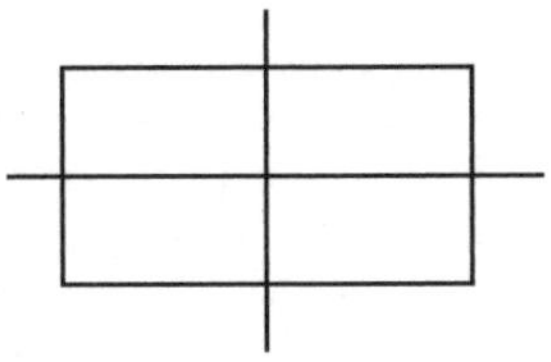 Award **ONE** mark for at least one correct line of symmetry and one or more incorrect lines, e.g. 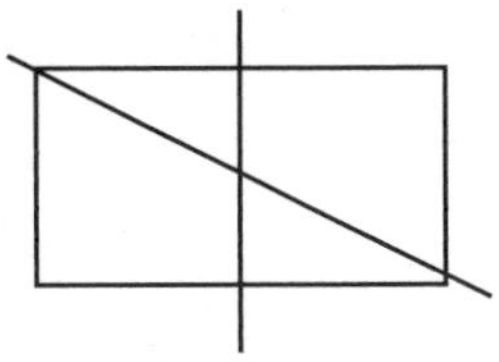 Award **ONE** mark for one correct line of symmetry and no other lines.	Up to 2	
4	Award **TWO** marks for the correct answer of 1,059. If the answer is incorrect, award **ONE** mark for an appropriate method, e.g. • 6,214 + 15,976 = 22,190 23,249 – 22,190 **OR** • 23,249 – 15,976 – 6, 214	Up to 2	Answer is not needed for the award of **ONE** mark.
5	6 –4	1 1	Accept –6
6	Award **TWO** marks for the correct answer of $\frac{1}{6}$ If the answer is incorrect, award **ONE** mark for an appropriate method, e.g. • $\frac{1}{2}+\frac{1}{3}=\frac{3}{6}+\frac{2}{6}=\frac{5}{6}$ $1-\frac{5}{6}=\frac{1}{6}$ **OR** • $1-\frac{1}{2}-\frac{1}{3}$ **OR** 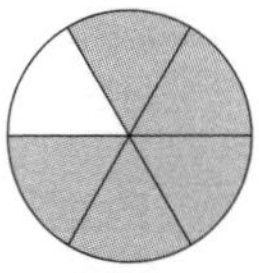 •	Up to 2	Accept equivalent fractions, or an exact decimal equivalent $0.1\dot{6}$ Award **ONE** mark for an answer between 0.16 and 0.17 Answer is not needed for the award of **ONE** mark.
	60° (360÷6)	1	Award **ONE** mark for a correct method of calculating a fraction of 360°.

Question	Answer	Mark	Additional guidance
Test 2			
1	Exactly seven months have more than 30 days. ✓	1	
2	Award **TWO** marks for the correct answer of 170 metres. If the answer is incorrect, award **ONE** mark for an appropriate method, e.g. • 50 + 35 + 35 + 50 **OR** • 100 + 70	Up to 2	Answer is not needed for the award of **ONE** mark.
3	The shape has moved **3** squares to the left and **5** squares up.	1	Both numbers must be corrrect for **ONE** mark.
4	**16** – **5** = 11	1	Numbers must be in the correct order. Do not accept $\mathbf{4^2}$ – **5** = 11
5	Award **TWO** marks for the correct answer of 28,610 If the answer is incorrect, award **ONE** mark for an appropriate method, e.g. • 23,050 – 2480 = 20,570 20,570 + 8,040 = 28,610 **OR** • 23,050 – 2,480 + 8,040	Up to 2	Answer is not needed for the award of **ONE** mark.
6	Award **TWO** marks for the correct answer of £1.20 If the answer is incorrect, award **ONE** mark for an appropriate method, e.g. • 1 + 1 + 1 + 1 + 2 = 6 6 ÷ 5	Up to 2	Answer is not needed for the award of **ONE** mark.
7	Award **THREE** marks for the correct answer of 36p If the answer is incorrect, award **ONE** mark for an appropriate method, e.g. • 19.88 ÷ 4 = £4.97 £4.97 ÷ 14 = 35.5p Award **ONE** mark for correct rounding of a decimal answer to the nearest penny.	Up to 3	Accept £0.36 Accept use of short or long division for 4.97 ÷ 14 Answer is not needed for the award of **TWO** marks.
Test 3			
1	Award **TWO** marks for the correct answer of 1,500 cars. If the answer is incorrect, award **ONE** mark for stating 1,500 red cars and/or 3,000 silver cars.	Up to 2	Answer is not needed for the award of **ONE** mark.
2	Award **TWO** marks for three boxes completed as shown: 24,390 24,400 24,000 Award **ONE** mark for any two of these boxes completed correctly.	Up to 2	
3	$\frac{1}{11}$	1	
4	Award **TWO** marks for correct answer of Yes and calculation 37 × 16 = 592 **OR** 500 ÷ 16 = 31.25 If the answer is incorrect, award **ONE** mark for correct calculation 37 × 16 = 592 **OR** 500 ÷ 16 = 31.25	Up to 2	Answer is not needed for the award of **ONE** mark. No marks can be awarded for Yes without the calculation.

<table>
<tr><th>Question</th><th>Answer</th><th>Mark</th><th>Additional guidance</th></tr>
<tr><td>5</td><td>Award THREE marks for correct answer of 28 cm
3 cm
4 cm
6 cm
5 cm
2 cm
8 cm 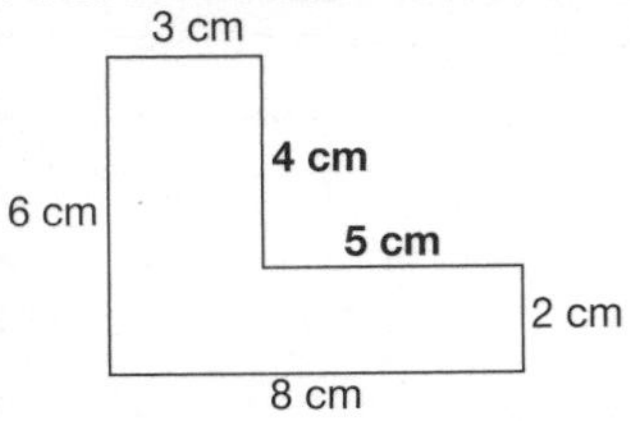
3 + 4 + 5 + 2 + 8 + 6 = 28 cm
Award TWO marks for incorrect answer to correct calculation 3 + 4 + 5 + 2 + 8 + 6
OR
Award TWO marks for correct answer to correct calculation using the four lengths given and one correct length (4 cm or 5 cm) found.
OR
Award ONE mark for one or both missing lengths 4 cm and 5 cm found.</td><td>Up to 3</td><td>Answer is not needed for the award of TWO marks or ONE mark.</td></tr>
<tr><td>6</td><td>£2.50</td><td>1</td><td>Accept £2.50p
Do not accept £2.5 or £250 or £250p</td></tr>
<tr><td colspan="4">Test 4</td></tr>
<tr><td>1</td><td>Award ONE mark for three correct answers, as shown:
<table>
<tr><td>4</td><td>×</td><td>12</td><td>=</td><td>48</td></tr>
<tr><td>×</td><td></td><td>×</td><td></td><td></td></tr>
<tr><td>7</td><td>×</td><td>6</td><td>=</td><td>42</td></tr>
<tr><td>=</td><td></td><td>=</td><td></td><td></td></tr>
<tr><td>28</td><td></td><td>72</td><td></td><td></td></tr>
</table></td><td>1</td><td>All numbers must be correct for ONE mark.</td></tr>
<tr><td>2</td><td>Award TWO marks for three angles labelled as shown (one each of R, A, O):
R O
A
R O
Award ONE mark for two angles labelled as shown (R and A, R and O, or O and A).</td><td>Up to 2</td><td>Do not award two marks if any angles are labelled incorrectly.</td></tr>
<tr><td>3</td><td>378 OR 738</td><td>1</td><td></td></tr>
<tr><td>4</td><td>18 minutes
Between Bute Road and Hill Street</td><td>1
1</td><td></td></tr>
<tr><td>5</td><td>Award ONE mark for the five numbers matched correctly, as shown:
<table>
<tr><td> 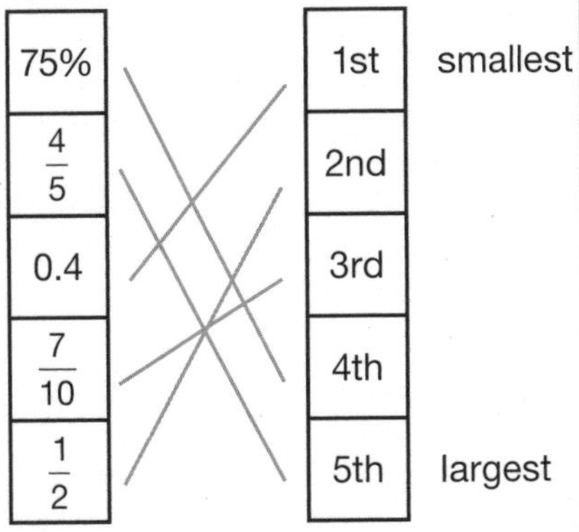75%</td><td>1st</td><td>smallest</td></tr>
<tr><td>$\frac{4}{5}$</td><td>2nd</td><td></td></tr>
<tr><td>0.4</td><td>3rd</td><td></td></tr>
<tr><td>$\frac{7}{10}$</td><td>4th</td><td></td></tr>
<tr><td>$\frac{1}{2}$</td><td>5th</td><td>largest</td></tr>
</table></td><td>1</td><td>Lines need not touch the numbers and ordinals, provided the intention is clear.

Do not accept any number which has been matched to more than one ordinal.</td></tr>
</table>

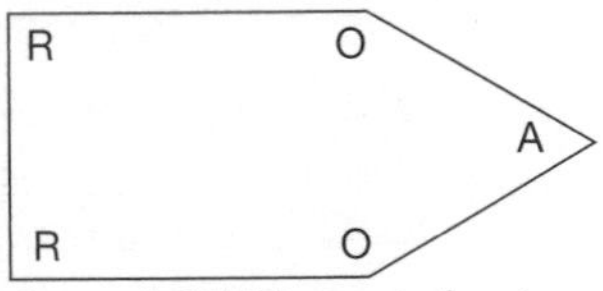

Question	Answer	Mark	Additional guidance
6	Award **THREE** marks for the correct answer of 4 cm If the answer is incorrect, award **TWO** marks for: • mention of 56 and 60 as evidence of 17 + 11 + 17 + 11 and 4 × 15 **OR** • an appropriate method with no more than one arithmetic error, e.g. • 4 × 15 = 60 17 + 13 (*error*) + 17 + 13 (*error*) = 60 60 – 60 = 0 Award **ONE** mark for an appropriate method.	Up to 3	Answer is not needed for the award of **ONE** mark. No marks are awarded if there is more than one misread of a number. **TWO** marks are awarded for an appropriate method using the misread number if it is followed through correctly to a final answer. **ONE** mark is awarded for an appropriate method using the misread number followed through correctly with no more than one arithmetic error.
Test 5			
1	Award **TWO** marks for all three numbers correct: **996** **936** **896** Award **ONE** mark for two numbers correct.	Up to 2	
2	Award **TWO** marks for the correct answer of 224 If the answer is incorrect, award **ONE** mark for an appropriate method, e.g. • 4 × 8 × 7	Up to 2	Answer is not needed for the award of **ONE** mark.
3	(4, 2)	1	
4	1 °C	1	
5	Award **THREE** marks for the correct answer of £72.41 If the answer is incorrect, award **TWO** marks for: • mention of £62.91 and £9.50 as evidence of 9 × £6.99 and 9 × 50p + £5 **OR** • an appropriate method with no more than one arithmetic error, e.g. 9 × £6.99 = £61.91 *(error)* £61.91 + £9.50 = £71.41 Award **ONE** mark for an appropriate method.	Up to 3	Accept £72.41p Do not accept £7241 or £7241p Answer is not needed for the award of **ONE** mark. No marks are awarded if there is more than one misread of a number. **TWO** marks are awarded for an appropriate method using the misread number if it is followed through correctly to a final answer. **ONE** mark is awarded for an appropriate method using the misread number followed through correctly with no more than one arithmetic error.
6	2,600 km ✓	1	
Test 6			
1	Award **TWO** marks for four masses in the order shown: 30 g, 200 g, 2000 g, 3 kg Award **ONE** mark for three of these masses in the correct order.	Up to 2	
2	C	1	

Question	Answer	Mark	Additional guidance
3	100	1	
4	Award **TWO** marks for the correct answers of 2, 3, 5 Award **ONE** mark for any two of 2, 3, 5	Up to 2	
5	Award **THREE** marks for the correct answer of 26 pages. If the answer is incorrect, award **TWO** marks for: • mention of 78 and 104 as evidence of $\frac{1}{4}$ of 312 and $\frac{1}{3}$ of 312. **OR** • an appropriate method with no more than one arithmetic error, e.g. $312 \div 4 = 78$ $312 \div 3 = 14$ *(error)* $78 - 14 = 64$ Award **ONE** mark for an appropriate method.	Up to 3	Answer is not needed for the award of **ONE** mark. No marks are awarded if there is more than one misread of a number. **TWO** marks are awarded for an appropriate method using the misread number if it is followed through correctly to a final answer. **ONE** mark is awarded for an appropriate method using the misread number followed through correctly with no more than one arithmetic error.
6	Award **ONE** mark for triangle A drawn as shown. Award **ONE** mark for translated triangle B drawn as shown. 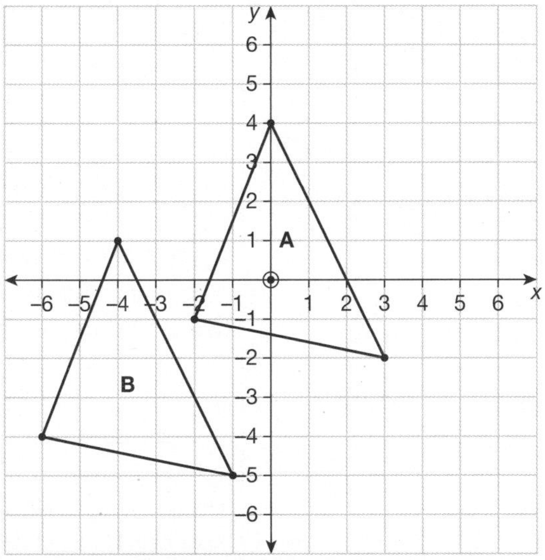	1 1	If triangle A is drawn incorrectly, award **ONE** mark for correct translation of triangle B.
Test 7			
1	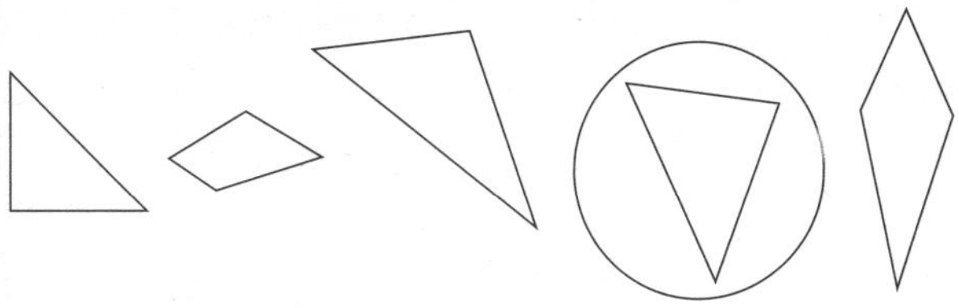	1	Accept correct answer clearly shown another way, e.g. ticking, underlining.
2	Award **TWO** marks for the correct answer of £1,295 If the answer is incorrect, award **ONE** mark for an appropriate method, e.g. • $542 + 1,740 = 2,282$ $2,282 - 987$	Up to 2	Do not accept £1295p Answer is not needed for the award of **ONE** mark.

Answers

Question	Answer	Mark	Additional guidance
3	$\frac{43}{9}$	1	Accept correct answer clearly shown another way, e.g. ticking, underlining
4	Line of symmetry Line of symmetry	1	Diagram does not need to be shaded.
5	21,250	1	
6	Award **TWO** marks for three boxes completed as shown: 234,820 234,800 235,000 Award **ONE** mark for any two of these boxes completed as shown.	Up to 2	
7	Award **THREE** marks for the correct answer of £420 If the answer is incorrect, award **TWO** marks for: • mention of 3,450 and 3,030 as evidence of 1,150 × 3 and 680 + 2,350 **OR** • an appropriate method with no more than one arithmetic error, e.g. 1,150 × 3 = 3,450 680 + 2,350 = 2,930 (*error*) 3,450 – 2,930 = 520 Award **ONE** mark for an appropriate method.	Up to 3	Answer is not needed for the award of **ONE** mark. No marks are awarded if there is more than one misread of a number. **TWO** marks are awarded for an appropriate method using the misread number if it is followed through correctly to a final answer. **ONE** mark is awarded for an appropriate method using the misread number followed through correctly with no more than one arithmetic error.
Test 8			
1	81 – 38 = 43 **OR** 81 – 43 = 38	1	All 6 digit cards must be completed correctly for **ONE** mark.
2	(5, 7)	1	

Answers

Question	Answer	Mark	Additional guidance
3	Award **TWO** marks for all four numbers placed as shown: Square numbers: 16, 4, 9; Both (intersection): 1, 64; Cube numbers: 27, 8 Award **ONE** mark for any three of these numbers placed as shown.	Up to 2	Do not accept numbers written in more than one region.
4	Award **TWO** marks for the correct answer of Jim with correct working shown: Jim £84 and Mel £91 If the answer is incorrect, award **ONE** mark for an appropriate method, e.g. • $8 \times 7 + 35$ 12×7	Up to 2	Answer is not needed for the award of **ONE** mark.
5	Award **TWO** marks for correct answer of $\frac{12}{25}$ If the answer is incorrect, award **ONE** mark for an appropriate method, e.g. • $\frac{48}{100}$ converted to a fraction but not simplified **OR** • $\frac{24}{50}$ simplified but not completely	Up to 2	
6	Award **THREE** marks for the correct answer of 37 cm^3 If the answer is incorrect, award **TWO** marks for: • mention of 27 and 64 as evidence of $3 \times 3 \times 3$ and $4 \times 4 \times 4$ **OR** • an appropriate method with no more than one arithmetic error, e.g. $3 \times 3 \times 3 = 9$ *(error)* $4 \times 4 \times 4 = 64$ $64 - 9 = 55$ Award **ONE** mark for an appropriate method.	Up to 3	Answer is not needed for the award of **ONE** mark. No marks are awarded if there is more than one misread of a number. **TWO** marks are awarded for an appropriate method using the misread number if it is followed through correctly to a final answer. **ONE** mark is awarded for an appropriate method using the misread number followed through correctly with no more than one arithmetic error.
Test 9			
1	$\frac{1}{2}$ $\frac{1}{11}$	1 1	
2	Award **ONE** mark for an explanation that 51×14 can be made by adding 51 to 663, e.g. • $51 \times 14 = 14 \times 51$ • $13 \times 51 = 663$, so add one more 51 to give you 14×51 • $663 + 51 = 714$	1	Do not accept an explanation that simply calculates $51 \times 14 = 714$ Do not accept vague, incomplete or incorrect explanations, e.g. 'You could add another 51'.

Answers

<table>
<tr><th>Question</th><th>Answer</th><th>Mark</th><th>Additional guidance</th></tr>
<tr><td>3</td><td>Award TWO marks for the correct answer of £0.49
If the answer is incorrect, award ONE mark for an appropriate method, e.g.
• 12 × 29p = £3.48
£3.48 – £2.99</td><td>Up to 2</td><td>Accept for ONE mark an answer of £49 OR £49p as evidence of an appropriate method.
Answer is not needed for the award of ONE mark.</td></tr>
<tr><td>4</td><td>56</td><td>1</td><td></td></tr>
<tr><td>5</td><td>Jack £14
Lucy £7</td><td>1
1</td><td>Award ONE mark for correct values (£14 and £7) matched to wrong names.</td></tr>
<tr><td>6</td><td>$x = 54°$
$y = 54°$
$z = 63° (\frac{1}{2}(180 - 54))$</td><td>1
1
1</td><td></td></tr>
<tr><td colspan="4">Test 10</td></tr>
<tr><td>1</td><td>£2.25</td><td>1</td><td>Accept £2.25p
Do not accept £225 or £225p</td></tr>
<tr><td>2</td><td>Award TWO marks for correct answer as shown:
4 7 2
× 8
3 7 7 6
Award ONE mark for one of the two digits correct.</td><td>Up to 2</td><td></td></tr>
<tr><td>3</td><td>10 cm
4 years old</td><td>1
1</td><td></td></tr>
<tr><td>4</td><td>16 hours</td><td>1</td><td></td></tr>
<tr><td>5</td><td>Award ONE mark for reflection drawn as shown:</td><td>1</td><td></td></tr>
<tr><td>6</td><td>Award THREE marks for all values completed as shown:
<table>
<tr><th>Angle</th><th>Favourite colour</th><th>Number of people</th></tr>
<tr><td>60°</td><td>Red</td><td>12</td></tr>
<tr><td>100°</td><td>Blue</td><td>20</td></tr>
<tr><td>130°</td><td>Green</td><td>26</td></tr>
<tr><td>70°</td><td>Yellow</td><td>14</td></tr>
<tr><td></td><td>Total</td><td>72</td></tr>
</table>
Award TWO marks for any four values completed as shown.
Award ONE mark for any three values completed as shown.</td><td>Up to 3</td><td></td></tr>
</table>

Answers

<table>
<tr><th>Question</th><th>Answer</th><th>Mark</th><th>Additional guidance</th></tr>
<tr><td colspan="4">Test 11</td></tr>
<tr><td>1</td><td>$\frac{2}{5} = \frac{4}{\mathbf{10}} = \frac{\mathbf{6}}{15}$</td><td>1</td><td>Both values must be correct for ONE mark.</td></tr>
<tr><td>2</td><td>Award TWO marks for table completed as shown:
<table><tr><th>Shape</th><th>Number of lines of symmetry</th></tr><tr><td>parallelogram</td><td>0</td></tr><tr><td>kite</td><td>1</td></tr><tr><td>rectangle</td><td>2</td></tr><tr><td>square</td><td>4</td></tr></table>Award ONE mark for any two shapes correctly placed.</td><td>Up to 2</td><td></td></tr>
<tr><td>3</td><td>1953</td><td>1</td><td></td></tr>
<tr><td>4</td><td>Award TWO marks for the correct answer of 24
If the answer is incorrect, award ONE mark for an appropriate method, e.g.
• 420 ÷ 68 and 280 ÷ 68
OR
• 420 ÷ 70 and 280 ÷ 70</td><td>Up to 2</td><td>Answer is not needed for the award of ONE mark.</td></tr>
<tr><td>5</td><td>Award TWO marks for all four given numbers placed correctly 4 times, as shown:
Prime numbers: 53
Multiples of 2: 54, 52, 56
Multiples of 3: 54, 51

If the answer is incorrect, award ONE mark for three of the given numbers all placed completely correctly.</td><td>1</td><td>Accept the numbers in any order. Ignore any additional numbers not given in the question.</td></tr>
<tr><td>6</td><td>Award THREE marks for the correct answer of £16.14
If the answer is incorrect, award TWO marks for:
• £16.13
OR
• mention of £4.40 or £48.40 as evidence of 10% calculated correctly and 16.13… as evidence of £48.40 ÷ 3
OR
• an appropriate method with no more than one arithmetic error, e.g.
£44 + 10% = £48 (error)
Award ONE mark for an appropriate method.</td><td>Up to 3</td><td>Accept £16.14p
Do not accept £1614 or £1614p
Answer is not needed for the award of ONE mark.
No marks are awarded if there is more than one misread of a number.
TWO marks will be awarded for an appropriate method using the misread number if it is followed through correctly to a final answer.
ONE mark will be awarded for an appropriate method using the misread number followed through correctly with no more than one arithmetic error.</td></tr>
<tr><td colspan="4">Test 12</td></tr>
<tr><td>1</td><td>3207 OR 3,207

Four thousand nine hundred and ninety four</td><td>1

1</td><td></td></tr>
<tr><td>2</td><td>$\frac{3}{9}$ OR $\frac{1}{3}$</td><td>1</td><td></td></tr>
</table>

<table>
<tr><th>Question</th><th>Answer</th><th>Mark</th><th>Additional guidance</th></tr>
<tr><td>3</td><td>(0.327) and (0.673)</td><td>1</td><td>Both numbers are needed for ONE mark.
Accept correct answers clearly shown another way, e.g. ticking, underlining.</td></tr>
<tr><td>4</td><td>Award THREE marks for the correct answer of £165
If the answer is incorrect, award TWO marks for:
• mention of 86.4 and 10.8 as evidence of 12 × 7.2 and 86.4 ÷ 8
OR
• an appropriate method with no more than one arithmetic error, e.g.
12 × 7.2 = 86.4
86.4 ÷ 8 = 18 (error)
Award ONE mark for an appropriate method.</td><td>Up to 3</td><td>Do not accept £165p
Answer is not needed for the award of ONE mark.
No marks are awarded if there is more than one misread of a number.
TWO marks are awarded for an appropriate method using the misread number if it is followed through correctly to a final answer.
ONE mark is awarded for an appropriate method using the misread number followed through correctly with no more than one arithmetic error.</td></tr>
<tr><td>5</td><td>33
6</td><td>1
1</td><td></td></tr>
<tr><td>6</td><td>Award TWO marks for Yes, with correct working as shown:
300 ÷ 4 = 75 ml
10 × 75 ml = 750 ml
750 ml < 1 litre
OR
300 ml for 4 people
150 ml for 2 people
300 + 300 + 150 ml = 750 ml for 10 people
750 ml < 1 litre
Award ONE mark for an appropriate method with no more than one arithmetic error.</td><td>Up to 2</td><td>Answer is not needed for the award of ONE mark.</td></tr>
<tr><td colspan="4">Test 13</td></tr>
<tr><td>1</td><td>mint and lemon OR lemon and mint
AND
lemon and toffee OR toffee and lemon</td><td>1</td><td>Both combinations are needed for ONE mark.</td></tr>
<tr><td>2</td><td>100</td><td>1</td><td></td></tr>
<tr><td>3</td><td>$\frac{2}{5}$ > 25%</td><td>1</td><td></td></tr>
<tr><td>4</td><td>Award TWO marks for the table completed as shown:
Award ONE mark for two of the missing numbers completed as shown.
<table><tr><th>Number</th><th>10,000 more</th></tr><tr><td>84,000</td><td>94,000</td></tr><tr><td>165</td><td>10,165</td></tr><tr><td>243,000</td><td>253,000</td></tr><tr><td>4,750</td><td>14,750</td></tr></table></td><td>Up to 2</td><td></td></tr>
</table>

Question	Answer	Mark	Additional guidance
5	E	1	
6	✓	1	Accept correct answer clearly shown another way, e.g. circling, underlining.
7	100 minutes	1	
	Award **TWO** marks for the correct answer of 4.10 pm If the answer is incorrect, award **ONE** mark for: • mention of 140 minutes or 2 hours 20 minutes as evidence of $3 \times 40 + 20$ **OR** • an appropriate method, e.g. $3 \times 40 + 20$ 2 hours 20 minutes before 6.30 pm	Up to 2	Answer is not needed for the award of **ONE** mark.
Test 14			
1	32	1	
	21	1	
2	$73 \times 80 =$ **5,840** $73 \times$ **0.8** $= 58.4$	1 1	
3	1 minute 50 seconds	1	
4	Award **ONE** mark for clear explanation with calculation to show that the third angle is 90°, which is a right angle, e.g. • 25° + 65° = 90° The other angle in the triangle is 180° – 90°= 90°	1	Do not award the mark for 'the other angle is a right angle' if it is not backed up with a calculation.
5	(triangle) costs **£4**	1	
	(trapezium) costs **£7**	1	
6	Award **TWO** marks for the correct answer of 5 with some working out shown. If the answer is incorrect, award **ONE** mark for: • mention of 5.625 as evidence of $5400 \div 960$ or $5.4 \div 0.96$ **OR** • an appropriate method.	Up to 2	Answer is not needed for the award of **ONE** mark.
	Award **ONE** mark for the correct answer of 600 mm	1	Accept 0.6 m or 60 cm

Score Chart

You have completed all the tests! Now write your scores in the score chart below.

Test	My Score
Test 1	/12
Test 2	/12
Test 3	/11
Test 4	/10
Test 5	/10
Test 6	/11
Test 7	/11
Test 8	/11
Test 9	/11
Test 10	/10
Test 11	/11
Test 12	/11
Test 13	/10
Test 14	/11
Total	**/152**

How did you do?